WORKSHEETS
WITH THE MATH COACH

GEX, INCORPORATED

BEGINNING ALGEBRA
EIGHTH EDITION

John Tobey
North Shore Community College

Jeffrey Slater
North Shore Community College

Jamie Blair
Orange Coast College

Jennifer Crawford
Normandale Community College

PEARSON

Boston Columbus Indianapolis New York San Francisco Upper Saddle River
Amsterdam Cape Town Dubai London Madrid Milan Munich Paris Montreal Toronto
Delhi Mexico City Sao Paulo Sydney Hong Kong Seoul Singapore Taipei Tokyo

Copyright © 2013, 2010, 2006 Pearson Education, Inc.
Publishing as Pearson, 75 Arlington Street, Boston, MA 02116.

ISBN-13: 978-0-321-76980-0
ISBN-10: 0-321-76980-5

4 5 6   V031   16 15 14 13

www.pearsonhighered.com

PEARSON

Worksheets with the Math Coach

Beginning Algebra, Eighth Edition

Table of Contents

Name: _____   Date: _____
Instructor: _____   Section: _____

Chapter 0 A Brief Review of Arithmetic Skills
0.1 Simplifying Fractions

Vocabulary
whole numbers • fractions • numerator • denominator • numerals
simplest form • reduced form • simplifying • reducing • natural numbers
counting numbers • factor • prime numbers • lowest terms • multiplicative identity
proper fraction • improper fraction • mixed number

1. The whole numbers 1, 2, 3, 4, 5, 6, 7, 8, and 9 are called _____ or counting numbers.

2. In the fraction $\dfrac{1}{25}$, 1 is the numerator and 25 is the _____.

3. When we obtain an equivalent fraction in simplest form (or reduced form), the new simplified fraction is said to be in _____.

4. The natural number factors of _____ are 1 and themselves.

Example	**Student Practice**
1. Simplify each fraction.	**2.** Simplify each fraction.
(a) $\dfrac{14}{21}$	(a) $\dfrac{33}{39}$
First factor 14 and 21, $\dfrac{14}{21} = \dfrac{7 \times 2}{7 \times 3}$.	
Now divide the numerator and denominator by 7.	
$\dfrac{14}{21} = \dfrac{\cancel{7} \times 2}{\cancel{7} \times 3} = \dfrac{2}{3}$	
(b) $\dfrac{20}{70}$	(b) $\dfrac{126}{210}$
Factor 20 and 70, $\dfrac{20}{70} = \dfrac{2 \times 2 \times 5}{7 \times 2 \times 5}$.	
Now divide the numerator and denominator by both 2 and 5.	
$\dfrac{20}{70} = \dfrac{2 \times \cancel{2} \times \cancel{5}}{7 \times \cancel{2} \times \cancel{5}} = \dfrac{2}{7}$	

Vocabulary Answers: 1. natural numbers 2. denominator 3. lowest terms 4. prime numbers

Example	Student Practice
3. Simplify the fraction $\dfrac{7}{21}$. First factor 7 and 21, $\dfrac{7}{21} = \dfrac{7 \times 1}{7 \times 3}$. Now divide the numerator and denominator by 7. $\dfrac{7}{21} = \dfrac{\cancel{7} \times 1}{\cancel{7} \times 3} = \dfrac{1}{3}$ Notice that all the prime numbers in the numerator divided out. When this happens, we must remember that 1 is left in the numerator.	**4.** Simplify the fraction $\dfrac{13}{169}$.
5. Simplify the fraction $\dfrac{70}{10}$. $\dfrac{70}{10} = \dfrac{7 \times \cancel{5} \times \cancel{2}}{\cancel{5} \times \cancel{2} \times 1} = 7$ Notice that all the prime numbers in the denominator divided out. When this happens, we do not need to leave 1 in the denominator because the answer is a whole number.	**6.** Simplify the fraction $\dfrac{423}{47}$.
7. Cindy got 48 out of 56 questions correct on a test. Write this as a fraction in simplest form. Express as a fraction the number of correct responses out of the total number of questions on the test. 48 out of 56 $\rightarrow \dfrac{48}{56}$ Express this fraction in simplest form. Factor 48 and 56, then divide out common factors. $\dfrac{48}{56} = \dfrac{6 \times \cancel{8}}{7 \times \cancel{8}} = \dfrac{6}{7}$ Cindy answers the questions correctly $\dfrac{6}{7}$ of the time.	**8.** Last June, Jacob went to the gym 18 out of the 30 days. Write this as a fraction in simplest form.

2

Example	Student Practice
9. Change $\dfrac{7}{4}$ to a mixed number or to a whole number.	**10.** Change $\dfrac{63}{7}$ to a mixed number or to a whole number.

Divide the denominator into the numerator.

$$\frac{7}{4} = 7 \div 4$$

$$\begin{array}{r} 1 \\ 4\overline{)7} \\ \underline{4} \\ 3 \quad \text{Remainder} \end{array}$$

The quotient, 1, is the whole-number part of the mixed number. The remainder from the division, 3, is the numerator of the fraction. The denominator of the fraction remains unchanged.

Thus, $\dfrac{7}{4} = 1\dfrac{3}{4}$.

11. Change $3\dfrac{1}{7}$ to an improper fraction.	**12.** Change $6\dfrac{5}{12}$ to an improper fraction

To change a mixed number to an improper fraction, multiply the whole number by the denominator. Add this to the numerator. The result is the new numerator. The denominator does not change.

$$3\frac{1}{7} = \frac{(3 \times 7)+1}{7} = \frac{21+1}{7} = \frac{22}{7}$$

Thus, $3\dfrac{1}{7} = \dfrac{22}{7}$.

3

Example	Student Practice
13. Find the missing number.	**14.** Find the missing number.

13. Find the missing number.

(a) $\dfrac{3}{5} = \dfrac{?}{25}$

A fraction can be changed to an equivalent fraction with a different denominator by multiplying both numerator and denominator by the same number. Observe that we need to multiply the denominator by 5 to obtain 25. So we multiply the numerator 3 by 5 also.

$$\dfrac{3\times5}{5\times5} = \dfrac{15}{25}$$

The desired number is 15.

(b) $\dfrac{2}{9} = \dfrac{?}{36}$

Observe that $9\times4 = 36$. We need to multiply the numerator by 4 to get the new numerator.

$$\dfrac{2\times4}{9\times4} = \dfrac{8}{36}$$

The desired number is 8.

14. Find the missing number.

(a) $\dfrac{1}{9} = \dfrac{?}{99}$

(b) $\dfrac{4}{7} = \dfrac{?}{35}$

Extra Practice

1. Simplify the fraction $\dfrac{60}{15}$.

2. Change $\dfrac{556}{10}$ to a mixed number.

3. Change $1\dfrac{12}{17}$ to an improper fraction.

4. Find the missing numerator, $\dfrac{8}{15} = \dfrac{?}{120}$.

Concept Check
Explain in your own words how to change a mixed number to an improper fraction.

4

Copyright © 2013 Pearson Education, Inc.

Chapter 0 A Brief Review of Arithmetic Skills
0.2 Adding and Subtracting Fractions

Vocabulary

fraction • numerator • denominator • common denominator
least common denominator • prime numbers • mixed numbers • perimeter

1. The _____ of two or more fractions is the smallest whole number that is exactly divisible by each denominator of the fractions.

2. Before you can add or subtract fractions, they must have the same _____.

3. The distance around a polygon is called the _____.

4. When adding or subtracting _____ first change them to improper fractions.

Example	**Student Practice**
1. Add the fractions. Simplify your answer whenever possible.	**2.** Add the fractions. Simplify your answer whenever possible.
(a) $\dfrac{5}{7}+\dfrac{1}{7}$	**(a)** $\dfrac{1}{4}+\dfrac{3}{4}$
Add the numerators and keep the denominator the same. $$\frac{5}{7}+\frac{1}{7}=\frac{5+1}{7}=\frac{6}{7}$$ This fraction cannot be simplified further.	
(b) $\dfrac{1}{8}+\dfrac{3}{8}+\dfrac{2}{8}$	**(b)** $\dfrac{7}{10}+\dfrac{5}{10}+\dfrac{3}{10}$
Add the numerators and keep the denominator the same. $$\frac{1}{8}+\frac{3}{8}+\frac{2}{8}=\frac{1+3+2}{8}=\frac{6}{8}$$ Now simplify. $$\frac{6}{8}=\frac{3}{4}$$	

Vocabulary Answers: 1. least common denominator 2. denominator 3. perimeter 4. mixed numbers

Example	Student Practice
3. Subtract the fractions $\dfrac{9}{11} - \dfrac{2}{11}$. Simplify your answer if possible. Subtract the numerators and keep the denominator the same. $\dfrac{9}{11} - \dfrac{2}{11} = \dfrac{9-2}{11} = \dfrac{7}{11}$ This fraction cannot be simplified further.	**4.** Subtract the fractions $\dfrac{7}{8} - \dfrac{5}{8}$. Simplify your answer if possible.
5. Find the LCD of $\dfrac{5}{6}$ and $\dfrac{1}{15}$ using the prime factor method. Write each denominator as the product of prime factors. The LCD is a product containing each different prime factor. $\quad 6 = 2 \cdot 3$ $\quad 15 = \downarrow \; 3 \cdot 5$ $\qquad \downarrow \; \downarrow \; \downarrow$ $\text{LCD} = 2 \cdot 3 \cdot 5$ The different factors are 2, 3, and 5, and each factor appears at most once in any one denominator. Multiply the factors to find the LCD. $\text{LCD} = 2 \cdot 3 \cdot 5 = 30$	**6.** Find the LCD of $\dfrac{1}{10}$ and $\dfrac{6}{35}$ using the prime factor method.
7. Find the LCD of $\dfrac{5}{12}$, $\dfrac{1}{15}$, and $\dfrac{7}{30}$. Write each denominator as the product of prime factors. The LCD is a product containing each different factor, with the factor 2 appearing twice since it occurs twice in the factorization of 12. $\quad 12 = 2 \cdot 2 \cdot 3$ $\quad 15 = \downarrow \quad\;\; 3 \cdot 5$ $\qquad\;\; \downarrow$ $\quad 30 = \downarrow \; 2 \cdot 3 \cdot 5$ $\qquad\;\; \downarrow \; \downarrow \; \downarrow$ $\text{LCD} = 2 \cdot 2 \cdot 3 \cdot 5 = 60$	**8.** Find the LCD of $\dfrac{11}{18}$, $\dfrac{2}{21}$, and $\dfrac{13}{42}$.

6

Example	Student Practice

9. Combine. $\dfrac{1}{5}+\dfrac{1}{6}-\dfrac{3}{10}$

First find the LCD.

$$5 = 5$$
$$6 = \quad 2\cdot 3$$
$$10 = 5\cdot 2 \quad\downarrow$$
$$\downarrow\ \downarrow\ \downarrow$$
$$LCD = 5\cdot 2\cdot 3 = 30$$

Now we change $\dfrac{1}{5}$, $\dfrac{1}{6}$, and $\dfrac{3}{10}$ to equivalent fractions that have the LCD for a denominator.

$$\dfrac{1}{5}=\dfrac{?}{30}\qquad \dfrac{1\times 6}{5\times 6}=\dfrac{6}{30}$$

$$\dfrac{1}{6}=\dfrac{?}{30}\qquad \dfrac{1\times 5}{6\times 5}=\dfrac{5}{30}$$

$$\dfrac{3}{10}=\dfrac{?}{30}\qquad \dfrac{3\times 3}{10\times 3}=\dfrac{9}{30}$$

Combine the three fractions.

$$\dfrac{1}{5}+\dfrac{1}{6}-\dfrac{3}{10}=\dfrac{6}{30}+\dfrac{5}{30}-\dfrac{9}{30}$$

$$=\dfrac{6+5-9}{30}$$

$$=\dfrac{2}{30}$$

Now simplify.

$$\dfrac{2}{30}=\dfrac{1}{15}$$

Thus, $\dfrac{1}{5}+\dfrac{1}{6}-\dfrac{3}{10}=\dfrac{1}{15}$.

10. Combine. $\dfrac{1}{4}+\dfrac{13}{18}-\dfrac{5}{54}$

Example	Student Practice

11. Manuel is enclosing a triangular shaped exercise yard for his new dog. He wants to determine how many feet of fencing he will need. The sides of the yard measure $20\frac{3}{4}$ feet, $15\frac{1}{2}$ feet, and $18\frac{1}{8}$ feet. What is the perimeter of (the total distance around) the triangle?

Begin by drawing a picture. Find the perimeter by adding up the lengths of the three sides of the triangle. To add mixed numbers, change them to improper fractions then add.

$$20\frac{3}{4}+15\frac{1}{2}+18\frac{1}{8}=\frac{83}{4}+\frac{31}{2}+\frac{145}{8}$$

$$=\frac{166}{8}+\frac{124}{8}+\frac{145}{8}$$

$$=\frac{435}{8}=54\frac{3}{8}$$

He will need $54\frac{3}{8}$ feet of fencing.

12. Find the perimeter of the parallelogram shaped park below.

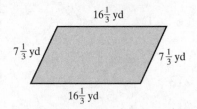

$16\frac{1}{3}$ yd

$7\frac{1}{3}$ yd $7\frac{1}{3}$ yd

$16\frac{1}{3}$ yd

Extra Practice

1. Find the LCD of $\frac{6}{25}$ and $\frac{99}{100}$. Do not combine the fractions; only find the LCD.

2. Combine $\frac{31}{45}-\frac{2}{15}$. Be sure to simplify your answer if possible.

3. Combine $\frac{7}{9}+2\frac{5}{12}$. Be sure to simplify your answer if possible.

4. Combine $6\frac{3}{11}-2\frac{17}{22}$. Be sure to simplify your answer if possible.

Concept Check

Explain how you would find the LCD of the fractions $\frac{4}{21}$ and $\frac{5}{18}$.

Chapter 0 A Brief Review of Arithmetic Skills
0.3 Multiplying and Dividing Fractions

Vocabulary
fractions • numerators • denominators • invert and multiply method
common factors • prime numbers • mixed numbers • complex fraction

1. A(n) _____ has one fraction in the numerator and one fraction in the denominator.

2. To multiply two fractions, multiply the _____ and multiply the denominators.

3. Use the _____ to divide two fractions.

4. When multiplying or dividing _____, first change them to improper fractions.

Example	Student Practice
1. Multiply.	**2.** Multiply.
(a) $\dfrac{3}{5} \times \dfrac{5}{7}$	(a) $\dfrac{2}{7} \times \dfrac{7}{17}$
Multiply the numerators and multiply the denominators. $$\dfrac{3}{5} \times \dfrac{5}{7} = \dfrac{3 \cdot 5}{5 \cdot 7}$$ Divide the numerator and denominator by 5 and simplify. $$\dfrac{3}{5} \times \dfrac{5}{7} = \dfrac{3 \cdot 5}{5 \cdot 7} = \dfrac{3 \cdot \cancel{5}^{1}}{\cancel{5}_{1} \cdot 7} = \dfrac{3}{7}$$	
(b) $\dfrac{15}{8} \times \dfrac{10}{27}$	(b) $\dfrac{55}{12} \times \dfrac{4}{45}$
If we factor each number, we can see the common factors. Remove common factors and multiply. $$\dfrac{15}{8} \times \dfrac{10}{27} = \dfrac{\cancel{3}^{1} \cdot 5}{2 \cdot 2 \cdot \cancel{2}_{1}} \times \dfrac{5 \cdot \cancel{2}^{1}}{\cancel{3}_{1} \cdot 3 \cdot 3} = \dfrac{25}{36}$$	

Vocabulary Answers: 1. complex fraction 2. numerators 3. invert and multiply method 4. mixed numbers

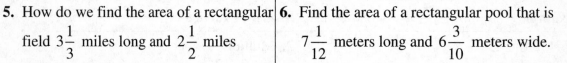

Example	Student Practice

3. Multiply $7 \times \dfrac{3}{5}$.

Write the whole number as a fraction whose denominator is 1.

$$7 \times \dfrac{3}{5} = \dfrac{7}{1} \times \dfrac{3}{5}$$

Follow the multiplication rule for fractions. Multiply numerators and multiply denominators.

$$7 \times \dfrac{3}{5} = \dfrac{7}{1} \times \dfrac{3}{5} = \dfrac{21}{5} \text{ or } 4\dfrac{1}{5}$$

4. Multiply $25 \times \dfrac{4}{5}$.

---

5. How do we find the area of a rectangular field $3\dfrac{1}{3}$ miles long and $2\dfrac{1}{2}$ miles wide?

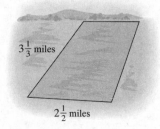

$3\frac{1}{3}$ miles

$2\frac{1}{2}$ miles

To find the area, we multiply length times width, $3\dfrac{1}{3} \times 2\dfrac{1}{2}$.

To multiply two mixed numbers, change them to improper fractions then follow the multiplication rule for fractions.

$$3\dfrac{1}{3} \times 2\dfrac{1}{2} = \dfrac{10}{3} \times \dfrac{5}{2} = \dfrac{\cancel{2} \cdot 5}{3} \times \dfrac{5}{\cancel{2}} = \dfrac{25}{3}$$

Write the answer as a mixed number.

The area is $8\dfrac{1}{3}$ square miles.

6. Find the area of a rectangular pool that is $7\dfrac{1}{12}$ meters long and $6\dfrac{3}{10}$ meters wide.

10

Example	Student Practice
7. Divide $\dfrac{2}{5} \div \dfrac{3}{10}$.	**8.** Divide $\dfrac{1}{9} \div \dfrac{5}{6}$.

7. Divide $\dfrac{2}{5} \div \dfrac{3}{10}$.

To divide two fractions, invert the second fraction (the divisor) and multiply.

$$\frac{2}{5} \div \frac{3}{10} = \frac{2}{5} \times \frac{10}{3}$$

Remove common factors and simplify.

$$\frac{2}{5} \div \frac{3}{10} = \frac{2}{5} \times \frac{10}{3}$$

$$= \frac{2}{\cancel{5}} \times \frac{\cancel{5} \cdot 2}{3} = \frac{4}{3} \text{ or } 1\frac{1}{3}$$

8. Divide $\dfrac{1}{9} \div \dfrac{5}{6}$.

---

9. Divide $\dfrac{1}{3} \div 2$.

Write the whole number as a fraction. Then invert the second fraction (the divisor) and multiply.

$$\frac{1}{3} \div 2 = \frac{1}{3} \div \frac{2}{1} = \frac{1}{3} \times \frac{1}{2} = \frac{1}{6}$$

10. Divide $7 \div \dfrac{1}{5}$.

---

11. Divide $\dfrac{\frac{3}{7}}{\frac{3}{5}}$.

This division is in the form of a complex fraction. Write this problem as a standard division first, then complete the problem using the rule for division.

$$\frac{\frac{3}{7}}{\frac{3}{5}} = \frac{3}{7} \div \frac{3}{5} = \frac{\cancel{3}}{7} \times \frac{5}{\cancel{3}} = \frac{5}{7}$$

12. Divide $\dfrac{\frac{7}{20}}{\frac{14}{15}}$.

Example	Student Practice
13. Divide $2\dfrac{1}{3} \div 3\dfrac{2}{3}$. To divide mixed numbers, change the mixed numbers to improper fractions and then use the rule for dividing fractions. $2\dfrac{1}{3} \div 3\dfrac{2}{3} = \dfrac{7}{3} \div \dfrac{11}{3} = \dfrac{7}{\cancel{3}} \times \dfrac{\cancel{3}}{11} = \dfrac{7}{11}$	**14.** Divide $\dfrac{10}{2\dfrac{4}{5}}$.

15. A chemist has 96 fluid ounces of a solution. She pours the solution into test tubes. Each test tube holds $\dfrac{3}{4}$ fluid ounces. How many test tubes can she fill?

Divide the total number of ounces by the number of ounces in each test tube.

$$96 \div \dfrac{3}{4} = \dfrac{96}{1} \div \dfrac{3}{4}$$

$$= \dfrac{96}{1} \times \dfrac{4}{3} = \dfrac{\cancel{3} \cdot 32}{1} \times \dfrac{4}{\cancel{3}} = 128$$

She will be able to fill 128 test tubes.

16. A graphic designer created 5 images for the evening news program in $6\dfrac{1}{4}$ hours. How many images did the designer create per hour?

Extra Practice

1. Multiply $\dfrac{3}{7} \times \dfrac{5}{12} \times \dfrac{8}{35}$. Simplify your answer if possible.

2. Multiply $7\dfrac{1}{2} \times 2\dfrac{2}{5}$. Simplify your answer if possible.

3. Divide $\dfrac{2}{3} \div \dfrac{4}{9}$. Simplify your answer if possible.

4. Mario cuts $59\dfrac{1}{2}$ inches of string into pieces that are exactly $4\dfrac{1}{4}$ inches long. How many pieces of string does Mario cut?

Concept Check

Explain the steps you would take to perform the calculation $3\dfrac{1}{4} \div 2\dfrac{1}{2}$.

Chapter 0 A Brief Review of Arithmetic Skills
0.4 Using Decimals

Vocabulary

decimal   •   decimal points   •   decimal places   •   terminal zeros   •   divisor
dividend   •   quotient   •   caret   •   fraction

1. A fraction whose denominator is 10, 100, and so on can be expressed as a _____.

2. Divide the dividend by the _____.

3. The number of digits to the right of the decimal point is also the number of
    _____.

4. To add or subtract decimals, write in column form and line up the _____.

Example	**Student Practice**
1. Write each of the following decimals as a fraction or mixed number. State the number of decimal places. Write out in words the way the number would be spoken. **(a)** 0.6 In fraction form, $0.6 = \dfrac{6}{10}$. 0.6 has 1 decimal place because it has 1 digit to the right of the decimal point. Use a place value chart if necessary to determine the word name. Written out, 0.6 is six-tenths. **(b)** 1.38 In fraction form, $1.38 = 1\dfrac{38}{100}$. 1.38 has 2 decimal places because it has 2 digits to the right of the decimal point. Use a place value chart if necessary to determine the word name. Written out, 1.38 is one and thirty-eight hundredths.	**2.** Write each of the following decimals as a fraction or mixed number. State the number of decimal places. Write out in words the way the number would be spoken. **(a)** 0.00009 **(b)** 4.025

Vocabulary Answers: 1. decimal 2. divisor 3. decimal places 4. decimal points

Example	Student Practice
3. Write $\dfrac{2}{11}$ as a decimal.	**4.** Write $\dfrac{13}{15}$ as a decimal.

$$11\overline{)2.0000}\;\;\;\dfrac{0.1818}{}$$

$$\underline{1\,1}$$
$$90$$
$$\underline{88}$$
$$20$$
$$\underline{11}$$
$$90$$
$$\underline{88}$$
$$2$$

Thus, $\dfrac{2}{11} = 0.1818\ldots = 0.\overline{18}$.

---

5. Write 0.138 as a fraction and simplify if possible.

To convert from a decimal to a fraction, write the decimal as a fraction with a denominator of 10, 100, 1000, and so on, and simplify the result.

Write 0.138 as a fraction with numerator 138 and denominator 1000. Then simplify.

$$0.138 = \dfrac{138}{1000} = \dfrac{69}{500}$$

6. Write 0.72 as a fraction and simplify if possible.

---

7. Subtract $127.32 - 38.48$.

To add or subtract decimals, write in column form and line up the decimal points. Then add or subtract the digits.

$$127.32$$
$$-\;\;\underline{38.48}$$
$$88.84$$

8. Add $21.16 + 2.04 + 13.91$.

14

Example	Student Practice

9. Multiply 2.56×0.003.

To multiply decimals, first multiply as with whole numbers. To determine the position of the decimal point, count the total number of decimal places in the two numbers being multiplied. This will determine the number of decimal places that should appear in the answer.

$$
\begin{array}{r}
2.56 \quad \text{(two decimal places)} \\
\times \ \ 0.003 \quad \text{(three decimal places)} \\
\hline
0.00768 \quad \text{(five decimal places)}
\end{array}
$$

Thus, $2.56 \times 0.003 = 0.00768$.

---

10. Multiply 0.85×0.0005.

---

11. Divide $16.2 \div 0.027$.

There are three decimal places in the divisor, so we move the decimal point three places to the right in the divisor and dividend and mark the new position by a caret.

Note that we must add two zeros to 16.2 in order to do this.

$$
0.027 _\wedge \overline{)16.200 _\wedge} \ \ \ \ ^{600.}
$$

Now perform the division as with whole numbers. The decimal point in the answer is directly above the caret.

$$
\begin{array}{r}
600. \\
0.027 _\wedge \overline{)16.200 _\wedge} \\
\underline{16 \ 2} \\
000
\end{array}
$$

Thus, $16.2 \div 0.027 = 600$.

---

12. Divide $25.2 \div 0.0003$.

Example	Student Practice
13. Multiply 0.0026×1000. When multiplying by 10, 100, 1000, and so on, for every zero in the multiplier, move the decimal point one place to the right. Since 1000 has 3 zeros, move the decimal point 3 places to the right. $0.0026 \times 1000 = 2.6$	**14.** Multiply 0.93×100.
15. Divide $0.0038 \div 100$. When dividing by 10, 100, 1000, and so on, for every zero in the divisor, move the decimal point one place to the left. Since 100 has 2 zeros, move the decimal point 2 places to the left. $0.0038 \div 100 = 0.000038$	**16.** Divide $7138.3 \div 10$.

Extra Practice

1. Write 32.082 as a fraction in simplified from. Write the value in words.

2. Subtract $14.03 - 7.8932$.

3. Divide $1.1592 \div 0.06$.

4. Multiply 16.785×100 by moving the decimal point.

Concept Check

Explain how you would place the decimal points when performing the calculation $0.252 \div 0.0035$.

Chapter 0 A Brief Review of Arithmetic Skills
0.5 Percents, Rounding, and Estimating

Vocabulary

decimal   •   decimal point   •   percent   •   % symbol   •   estimation   •   nonzero digit

1. A _____ is a fraction that has a denominator of 100.

2. To estimate by rounding, first round each number so that there is one _____.

3. _____ is the process of finding an approximate answer.

4. To change a percent to a decimal, move the _____ two places to the left and remove the % symbol.

Example	Student Practice
1. Change to a percent. (a) 0.0364 We move the decimal point two places to the right and add the % symbol. 0.0364 = 3.64% (b) 0.4 0.4 = 40%	**2.** Change to a percent. (a) 0.00019 (b) 0.6
3. Change to a percent. (a) 2.938 We move the decimal point two places to the right and add the % symbol. 2.938 = 293.8% (b) 4.5 4.5 = 450%	**4.** Change to a percent. (a) 5.95 (b) 7.9

Vocabulary Answers: 1. percent 2. nonzero digit 3. estimation 4. decimal point

Example	Student Practice

5. Change to a decimal.
   (a) 4%
     First we move the decimal point two places to the left. Then we remove the % symbol.
 	
     $4\% = 4\ .\ \% = 0.04$
              ↑
     The unwritten decimal point is understood to be here.
 	
   (b) 254.8%
     254.8% = 2.548
	6. Change to a decimal.
	   (a) 0.8%
	
	   (b) 101.2%

7. Find 182% of 12. To find the percent of a number, change the percent to a decimal and multiply the number by the decimal. 182% of $12 = 1.82 \times 12 = 21.84$	**8.** Find 5% of 639.

9. A store is having a sale of 35% off the retail price of all sofas. Melissa wants to buy a particular sofa that normally sells for $595.
   (a) How much will Melissa save if she buys the sofa on sale?
 	
     Find 35% of $595.
     $0.35 \times 595 = 208.25$
 	
     Thus, Melissa will save $208.25.
 	
   (b) What will the purchase price be if Melissa buys the sofa on sale?
 	
     The purchase price is the difference between the original price and the amount saved.
     $595.00 - 208.25 = 386.75$
 	
     If Melissa buys the sofa on sale, she will pay $386.75.
	10. A store is having a sale of 65% off the retail price of all boots. Marcus wants to buy a particular pair of boots that normally sells for $123.
	   (a) How much will Marcus save if he buys the boots on sale?
	
	   (b) What will the purchase price be if Marcus buys the boots on sale?

Example	Student Practice
11. What percent of 24 is 15?	**12.** What percent of 72 is 63?

11. What percent of 24 is 15?

First, write a fraction with the two numbers. The number after the word of is always the denominator, and the other number is the numerator.

$$\frac{15}{24}$$

Simplify the fraction.

$$\frac{15}{24} = \frac{5}{8}$$

Change the simplified fraction to a decimal.

$$\frac{5}{8} = 0.625$$

Express the decimal as a percent.
$0.625 = 62.5\%$

Thus, 62.5% of 24 is 15.

12. What percent of 72 is 63?

---

13. Marcia made 29 shots on goal during the last high school field hockey game. She actually scored a goal 8 times. What percent of her total shots were goals? Round your answer to the nearest whole percent.

We want to know what percent of 29 is 8. This relationship expressed as a fraction is $\frac{8}{29}$.

This fraction cannot be reduced. The next step is to write the fraction as a decimal.

$$\frac{8}{29} = 0.2758...$$

Change the decimal to a percent and round to the nearest whole number.
$0.2758... = 27.58...\% \approx 28\%$

Thus, Marcia scored a goal about 28% of the time she made a shot on goal.

14. In a shipment of 600 widgets, 38 widgets are warped. What percent of the total widgets are warped? Round your answer to the nearest whole percent.

Example	Student Practice

Example

15. The four walls of a college classroom are $22\frac{1}{4}$ feet long and $8\frac{3}{4}$ feet high. A painter needs to know the area of these four walls in square feet. Since paint is sold in gallons, an estimate will do. Use estimation by rounding to approximate the area of the four walls.

To estimate by rounding, round each number so that there is one nonzero digit. Round $22\frac{1}{4}$ feet to 20 feet.

Round $8\frac{3}{4}$ feet to 9 feet.

Multiply 20×9 to obtain an estimate of the area of one wall. Multiply $20\times9\times4$ to obtain an estimate of the area of all four walls.

$20\times9\times4 = 720$

Our estimate for the painter is 720 square feet of wall space.

Student Practice

16. A landscaper is to seed a lawn that is $47\frac{1}{2}$ meters long and $18\frac{1}{4}$ meters wide. The landscaper needs to know the area of the lawn in square meters to determine how much grass seed to buy. Use estimation by rounding to approximate the area of the lawn.

Extra Practice

1. Change 0.00576 to a percent.

2. Change 100% to a decimal.

3. 65 is what percent of 25?

4. Follow the principles of estimation to find an approximate value of $\frac{804}{39,500}$. Round each number so that there is one nonzero digit. Do not find the exact value.

Concept Check

Explain how you would change 0.0078 to a percent.

Name: _____   Date: _____
Instructor: _____   Section: _____

Chapter 0 A Brief Review of Arithmetic Skills
0.6 Using the Mathematics Blueprint for Problem Solving

Vocabulary

mathematics blueprint   •   area   •   percent   •   check   •   estimation

1. When solving real-life problems, use a(n) _____.

2. You can _____ your answer to real-life problems using estimation.

Example	Student Practice
1. Nancy and John want to install wall-to-wall carpeting in their living room. The floor of the rectangular living room is $11\frac{2}{3}$ feet wide and $19\frac{1}{2}$ feet long. How much will it cost if the carpet is $18.00 per square yard?	**2.** Kris wants to install wall-to-wall carpet in a bedroom. The floor of this rectangular room is $12\frac{1}{2}$ feet wide and $17\frac{1}{3}$ feet long. How much will it cost if the carpet is $27.00 per square yard?

Read the problem carefully and create a Mathematics Blueprint. Draw a picture if it will help.

Find the area of the floor.

$$11\frac{2}{3}\times19\frac{1}{2} = \frac{35}{3}\times\frac{39}{2} = \frac{455}{2} = 227\frac{1}{2}$$

A minimum of $227\frac{1}{2}$ square feet of carpet is needed. Dividing this value by 9 gives the area in square yards,

$$227\frac{1}{2}\div 9 = 25\frac{5}{18}.$$ Multiply $25\frac{5}{18}$ square yards by $18.00 per square yard to find the cost.

$$25\frac{5}{18}\times18 = \frac{455}{18}\times\frac{18}{1} = 455$$

The carpet will cost a minimum of $455.00 for this room. The check is left to the student.

Vocabulary Answers: 1. mathematics blueprint 2. check

Example	Student Practice

3. The following chart shows the 2012 sales of Micropower Computer Software for each of the four regions of the United States. Use the chart to answer the following questions (round all answers to the nearest whole percent).

Region of the U.S.	Number of Sales Personnel	Dollar Volume of Sales
Northeast	12	1,560,000
Southeast	18	4,300,000
Northwest	10	3,660,000
Southwest	15	3,720,000
Total	55	13,240,000

(a) What percent of the sales personnel are assigned to the Northeast?

Read the problem carefully and create a Mathematics Blueprint. You must calculate, "12 is what percent of 55?" Divide 12 by 55. Change the decimal to a percent and round.

$$\frac{12}{55} = 0.21818... = 21.818...\% \approx 22\%$$

Thus, about 22% of sales personnel are assigned to the Northeast.

(b) What percent of the volume of sales is attributed to the Northeast?

You must calculate, "1,560,000 is what percent of 13,240,000?"

$$\frac{1,560,000}{13,240,000} = \frac{156}{1324} \approx 0.1178 \approx 12\%$$

Thus, about 12% of the volume of sales is attributed to the Northeast.

4. Use the chart in example **3** to answer the following.

(a) What percent of sales personnel are assigned to the Northwest?

(b) What percent of the volume of sales is attributed to the Northwest?

(c) Of the Northeast and Northwest, which region has sales personnel that appear to be more effective in terms of the volume of sales?

Extra Practice

1. In a recent mayoral election in the town of Oceanside, Ms. Dempsey received 44% of the vote, Mr. Chan received 36%, and Ms. Tobin received 18%. A total of 6150 people voted. How many more people voted for Ms. Dempsey than Mr. Chan?

2. Ally would like to install linoleum on her kitchen floor. The floor of the rectangular kitchen is $16\frac{1}{2}$ feet wide and $22\frac{1}{2}$ feet long. How much will it cost to replace the floor if the linoleum costs $17.00 per square yard?

3. Nadine ran 2.2 kilometers on Monday. On each of the next four days, she increased her distance by 40% compared with the previous day. How many kilometers further did Nadine run on Thursday than on Wednesday? Round to the nearest tenth of a kilometer.

4. Karen earns $5200 per month, of which 35% is taken out in taxes and other withholding. How much does Karen take home every month after taxes and withholding?

Concept Check

Hank knows that 1 kilometer ≈ 0.62 mile. Explain how Hank could find out how many miles he traveled on a 12-kilometer trip to Mexico.

MATH COACH

Mastering the skills you need to do well on the test.

Watch the MATH COACH videos in MyMathLab®or on YouTube™ while you work the problems below. These helpful hints will help you avoid making common errors on test problems.

Subtracting Mixed Numbers—Problem 7

Subtract $3\dfrac{2}{3} - 2\dfrac{5}{6}$.

> **Helpful Hint:** First change the mixed numbers to improper fractions. Next find the LCD of the two denominators. Then change the fractions to an equivalent form with the LCD as the common denominator before subtracting.

Did you change $3\dfrac{2}{3}$ to $\dfrac{11}{3}$ and $2\dfrac{5}{6}$ to $\dfrac{17}{6}$?

Yes _____ No _____

If you answered No, stop and change the two mixed numbers to improper fractions.

Did you find the LCD to be 6? Yes _____ No _____

If you answered No, consider how to find the LCD of the two denominators. Once the fractions are written as equivalent fractions with 6 as the denominator, the two like fractions can be subtracted.

If you answered Problem 7 incorrectly, go back and rework the problem using these suggestions.

Dividing Mixed Numbers—Problem 10   Divide $5\dfrac{3}{8} \div 2\dfrac{3}{4}$.

> **Helpful Hint:** Be sure to change the mixed numbers to improper fractions before dividing.

Did you change $5\dfrac{3}{8}$ to $\dfrac{43}{8}$ and $2\dfrac{3}{4}$ to $\dfrac{11}{4}$ before doing any other steps? Yes _____ No _____

If you answered No, stop and change the two mixed numbers to improper fractions.

Next did you change the division to multiplication to obtain $\dfrac{43}{8} \times \dfrac{4}{11}$? Yes _____ No _____

If you answered No, stop and make this change.

Did you simplify the product? Yes _____ No _____

If you answered No, try dividing a 4 from the second numerator and first denominator before multiplying. The product will be an improper fraction that can be converted to a mixed number.

Now go back and rework the problem using these suggestions.

Dividing Decimals—Problem 17   Divide $12.88 \div 0.056$.

> **Helpful Hint:** Be careful as you move the decimal point in the divisor to the right. Make sure that the resulting divisor is an integer. Then move the decimal point in the dividend the same number of places to the right. Add zeros if necessary.

Did you move the decimal point in the divisor three places to the right to get 56?

Yes _____ No _____

If you answered No, stop and perform this step first.

Did you move the decimal point in the dividend three places to the right and add one zero to get 12880?

Yes _____ No _____

If you answered No, perform this step now. Be careful of calculation errors as you perform the division.

If you answered problem 17 incorrectly, go back and rework the problem using these suggestions.

Find the Missing Percent—Problem 22   39 is what percent of 650?

> **Helpful Hint:** Write a fraction with the two numbers. The number after the word "of" is always the denominator, and the other number is the numerator.

Did you write the fraction $\dfrac{39}{650}$?

Yes _____ No _____

If you answered No, stop and perform this step.

Did you simplify the fraction to $\dfrac{3}{50}$ before changing the fraction to a decimal?

Yes _____ No _____

If you answered No, consider that simplifying the fraction first makes the division step a little easier. Be sure to place the decimal point correctly in your quotient.

Did you change the quotient from a decimal to a percent?

Yes _____ No _____

If you answered No, stop and perform this final step.

Now go back and rework the problem using these suggestions.

Chapter 1 Real Numbers and Variables
1.1 Adding Real Numbers

Vocabulary
whole numbers • integers • rational numbers • irrational numbers • real numbers
number line • positive numbers • negative numbers • opposite numbers • absolute value
additive inverses

1. The _____ of a number is the distance between that number and zero on a
    number line.

2. _____, also called additive inverses, have the same magnitude but different signs
    and can be represented on a number line.

3. _____ are all the rational numbers and all the irrational numbers.

4. _____ are to the left of 0 on a number line.

Example	**Student Practice**
1. Classify as an integer, a rational number, an irrational number, and/or a real number.	**2.** Classify as an integer, a rational number, an irrational number, and/or a real number.
(a) 5	**(a)** $\sqrt{10}$
5 is an integer, a rational number, and a real number.	
(b) $-\dfrac{1}{3}$	**(b)** $\dfrac{1}{9}$
$-\dfrac{1}{3}$ is a rational number and a real number.	
(c) $\sqrt{2}$ is an irrational number and a real number.	**(c)** -4.25

Vocabulary Answers: 1. absolute value 2. opposite numbers 3. real numbers 4. negative numbers

Example	Student Practice
3. Use a real number to represent each situation.	**4.** Use a real number to represent each situation.
(a) A temperature of 128.6° F below zero is recorded at Vostok, Antarctica.	**(a)** A stock loss of 6.23 points
"below" is a keyword indicating that the number is negative.	
128.6° F below zero is -128.6.	**(b)** A temperature of 109.4° F in a desert
(b) The Himalayan peak K2 rises 29,064 feet above sea level.	
"above" is a keyword indicating that the number is positive.	**(c)** A population loss of 345
29,064 feet above sea level is $+29,064$.	
(c) The Dow gains 10.24 points.	**(d)** A 15 yard gain in a football game
A gain of 10.24 points is $+10.24$.	
(d) An oil drilling platform extends 328 feet below sea level.	
328 feet below sea level is -328.	
5. Find the additive inverse (that is, the opposite).	**6.** Find the additive inverse (that is, the opposite).
(a) -7	**(a)** $-\dfrac{7}{8}$
The opposite of -7 is $+7$.	
(b) $\dfrac{1}{4}$	**(b)** 30 feet below sea level
The opposite of $\dfrac{1}{4}$ is $-\dfrac{1}{4}$.	

Example	Student Practice
7. Find the absolute value.	**8.** Find the absolute value.

7. Find the absolute value.

   (a) $\left|-4.62\right|$

      $\left|-4.62\right| = 4.62$

   (b) $\left|\dfrac{3}{7}\right|$

      $\left|\dfrac{3}{7}\right| = \dfrac{3}{7}$

   (c) $\left|0\right|$

      $\left|0\right| = 0$

8. Find the absolute value.

   (a) $\left|5.34\right|$

   (b) $\left|-\sqrt{5}\right|$

   (c) $\left|\dfrac{0}{6}\right|$

---

9. Add. $14+6$

To add two numbers with the same sign, add the absolute values. Then use the common sign, $+$, in the answer.

$$14+16 = 30$$
$$14+16 = +30$$

10. Add. $-9+(-5)$

---

11. Add. $8+(-7)$

To add two numbers with different signs, find the difference between the two absolute values. The answer will have the sign of the number with the larger absolute value.

$$8-7 = 1$$
$$+8+(-7) = +1 \text{ or } 1$$

12. Add. $3+(-9)$

Example	Student Practice
13. Add. $-1.8 + 1.4 + (-2.6)$ We take the difference of 1.8 and 1.4 and use the sign of the number with the larger absolute value. Then add the result to -2.6. $-0.4 + (-2.6) = -3.0$	**14.** Add. $5.6 + (-2.34) + (-3.16)$
15. Add. $-8 + 3 + (-5) + (-2) + 6 + 5$ Add the three negative numbers and the three positive numbers separately, then add the two results. $\begin{array}{cc} -8 & +3 \\ -5 & +6 \\ \underline{-2} & \underline{+5} \\ -15 & +14 \end{array}$ Add the two results, $-15 + 14 = -1$.	**16.** Add. $-7 + 2 + 5 + (-6) + 1 + (-3)$

Extra Practice

1. Identify the following as a whole number, rational number, irrational number, integer, or real number. Remember, a number can be identified as more than one type.

    π

2. Find the absolute value of $-\dfrac{7}{8}$.

3. Use a real number to represent the situation.

    You owe your best friend $25.

4. Add. $57 + (-32) + 90 + (-100)$

Concept Check

Explain why when you add two negative numbers, you always obtain a negative number, but when you add one negative number and one positive number, you may obtain zero, a positive number, or a negative number.

Chapter 1 Real Numbers and Variables
1.2 Subtracting Real Numbers

Vocabulary
additive inverse property   •   subtract

1. To _____ real numbers, add the opposite of the second number to the first.

2. The _____ says that when you add two real numbers that are opposites of each other, you will obtain zero.

Example	Student Practice
1. Subtract. $6-(-2)$ Change the subtraction to addition and write the opposite of the second number. $6-(-2)=6+(+2)$ Then add the two real numbers with the same sign. $6+(+2)=8$	**2.** Subtract. $4-(-8)$
3. Subtract. $-8-(-6)$ Change the subtraction to addition and write the opposite of the second number. $-8-(-6)=-8+(+6)$ Then add the two real numbers with the same sign. $-8+(+6)=-2$	**4.** Subtract. $-15-(-9)$

Vocabulary Answers: 1. subtract 2. additive inverse property

Example	Student Practice
5. Subtract.	**6.** Subtract.

(a) $\dfrac{3}{7} - \dfrac{6}{7}$

Note that the problem has two fractions with the same denominator.

$$\dfrac{3}{7} - \dfrac{6}{7} = \dfrac{3}{7} + \left(-\dfrac{6}{7}\right)$$

$$= -\dfrac{3}{7}$$

(b) $-\dfrac{7}{18} - \left(-\dfrac{1}{9}\right)$

Change subtracting to adding the opposite and change $\dfrac{1}{9}$ to $\dfrac{2}{18}$ since the LCD $= 18$.

$$-\dfrac{7}{18} - \left(-\dfrac{1}{9}\right) = -\dfrac{7}{18} + \dfrac{1}{9}$$

$$= -\dfrac{7}{18} + \dfrac{2}{18}$$

$$= -\dfrac{5}{18}$$

Student Practice 6.

(a) $\dfrac{2}{9} - \dfrac{7}{9}$

(b) $-\dfrac{2}{5} - \left(-\dfrac{1}{4}\right)$

---

7. Subtract. $-5.2 - (-5.2)$

Change the subtraction problem to one of adding the opposite of the second number. Then, add two numbers with different signs.

$$-5.2 - (-5.2) = -5.2 + 5.2$$

$$= 0$$

8. Subtract. $-4.2 - (-4.2)$

32

Copyright © 2013 Pearson Education, Inc.

Example	Student Practice
9. Subtract.	**10.** Subtract.
(a) $-8 - 2$	**(a)** $-12 - 8$
To subtract, we add the opposite of the second number to the first.	
$-8 - 2 = -8 + (-2) = -10$	
(b) $23 - 28$	**(b)** $15 - 20$
In a similar fashion, we have the following.	
$23 - 28 = 23 + (-28) = -5$	
(c) $5 - (-3)$	**(c)** $14 - (-11)$
$5 - (-3) = 5 + 3 = 8$	
(d) $\dfrac{1}{4} - 8$	**(d)** $\dfrac{4}{5} - 6$
Write -8 as $-\dfrac{32}{4}$ since the LCD $= 4$.	
$\dfrac{1}{4} - 8 = \dfrac{1}{4} + (-8)$	
$\qquad = \dfrac{1}{4} + \left(-\dfrac{32}{4}\right)$	
$\qquad = -\dfrac{31}{4}$ or $-7\dfrac{3}{4}$	

Example	Student Practice
11. A satellite is recording radioactive emissions from nuclear waste buried 3 miles below sea level. The satellite orbits Earth at 98 miles above sea level. How far is the satellite from the nuclear waste?	**12.** A helicopter flies over a deep sea diver. The helicopter is 500 feet above sea level. The diver is 129 feet below sea level. How far is the helicopter from the diver?

We want to find the difference between +98 miles and −3 miles. This means we must subtract −3 from 98.

$$98 - (-3) = 93 + 3 = 101$$

The satellite is 101 miles from the nuclear waste.

Extra Practice

1. Subtract by adding the opposite.

$$(-80.09) - (-76.8)$$

2. Subtract by adding the opposite.

$$-\frac{11}{15} - \left(-\frac{1}{2}\right)$$

3. Change each subtraction operation to "adding the opposite." Then combine the numbers.

$$-30 + 12 - (-15) - 8$$

4. In the morning, Maxine began hiking uphill from a valley that is 22 feet below sea level. She ate her lunch on top of a hill whose altitude is 179 feet above sea level. Write an expression to represent the difference in altitude. How many feet did Maxine climb during her hike?

Concept Check

Explain the different results that are possible when you start with a negative number and then subtract a negative number.

Chapter 1 Real Numbers and Variables
1.3 Multiplying and Dividing Real Numbers

Vocabulary
multiplication • division • positive • negative • undefined

1. Division by zero is _____.

2. When you multiply or divide two numbers with different signs, you obtain a _____ number.

3. _____ can be indicated by the symbol ÷ or by the fraction bar −.

4. _____ is commutative and associative.

Example	Student Practice
1. Multiply.	**2.** Multiply.

1. Multiply.

(a) $\left(-\dfrac{5}{7}\right)\left(-\dfrac{2}{9}\right)$

When multiplying two numbers with the same sign, the result is a positive number.

$$\left(-\dfrac{5}{7}\right)\left(-\dfrac{2}{9}\right) = \dfrac{10}{63}$$

(b) $-4(8)$

When multiplying two numbers with different signs, the result is a negative number.

$$-4(8) = -32$$

2. Multiply.

(a) $(5)(4)$

(b) $\left(-\dfrac{5}{12}\right)(5)$

Vocabulary Answers: 1. undefined 2. negative 3. division 4. multiplication

Example	Student Practice
3. Multiply. **(a)** $\left(-\dfrac{1}{2}\right)(-1)(-4)$ Begin by multiplying the first two numbers. The signs are the same, so the answer is positive. Then multiply the remaining two numbers. The signs are different, so the answer is negative. $$\left(-\dfrac{1}{2}\right)(-1)(-4) = +\dfrac{1}{2}(-4) = -2$$ **(b)** $-2(-2)(-2)(-2)$ $$-2(-2)(-2)(-2) = +4(-2)(-2)$$ $$= -8(-2)$$ $$= +16 \text{ or } 16$$	**4.** Multiply. **(a)** $-6(-5.4)$ **(b)** $-3(-2)(-5)(-1)(-2)$
5. Divide. **(a)** $12 \div 4$ When dividing two numbers with the same sign, the result is a positive number. $12 \div 4 = 3$ **(b)** $\dfrac{-36}{18}$ When dividing two numbers with different signs, the result is a negative number. $$-\dfrac{36}{18} = -2$$	**6.** Divide. **(a)** $(-32) \div (-4)$ **(b)** $\dfrac{45}{-9}$

Example	Student Practice
7. Divide.	**8.** Divide.

7. Divide.

(a) $-36 \div 0.12$

When dividing two numbers with different signs, the result is a negative number. We then divide the absolute values.

$$0.12_\wedge \overline{)36.00_\wedge} \quad \begin{array}{c} 3\ 00. \\ \hline \underline{36} \\ 00 \end{array}$$

Thus $-36 \div 0.12 = -300$.

(b) $-2.4 \div (-0.6)$

$$0.6_\wedge \overline{)2.4_\wedge} \quad \begin{array}{c} 4. \\ \hline \underline{2\ 4} \end{array}$$

Thus $-2.4 \div (-0.6) = 4$.

8. Divide.

(a) $-121 \div (-0.11)$

(b) $-0.54 \div 0.9$

---

9. Divide. $-\dfrac{12}{5} \div \dfrac{2}{3}$

We invert the second fraction and multiply by the first fraction. The answer is negative since the two numbers divided have different signs.

$$-\frac{12}{5} \div \frac{2}{3} = \left(-\frac{12}{5}\right)\left(\frac{3}{2}\right)$$

$$= \left(-\frac{\overset{6}{\cancel{12}}}{5}\right)\left(\frac{3}{\underset{1}{\cancel{2}}}\right)$$

$$= -\frac{18}{5} \text{ or } -3\frac{3}{5}$$

10. Divide. $-\dfrac{2}{9} \div \left(-\dfrac{8}{15}\right)$

Example	Student Practice
11. Divide. $\dfrac{-\dfrac{2}{3}}{-\dfrac{7}{13}}$	**12.** Divide. $\dfrac{\dfrac{32}{4}}{-\dfrac{4}{5}}$

$$\dfrac{-\dfrac{2}{3}}{-\dfrac{7}{13}} = -\dfrac{2}{3} \div \left(-\dfrac{7}{13}\right)$$

$$= -\dfrac{2}{3}\left(-\dfrac{13}{7}\right)$$

$$= \dfrac{26}{21} \text{ or } 1\dfrac{5}{21}$$

Extra Practice

1. Multiply. Be sure to write your answer in the simplest form.

   $(15.8)(-29.3)$

2. Multiply. You may want to determine the sign of the product before you multiply.

   $(2.5)(-0.4)(-3.2)(5)$

3. Divide.

   $87.5 \div (-0.5)$

4. Divide.

   $-\dfrac{11}{16} \div \dfrac{33}{40}$

Concept Check

Explain how you can determine the sign of the answer if you multiply several negative numbers.

Name: _____   Date: _____
Instructor: _____   Section: _____

Chapter 1 Real Numbers and Variables
1.4 Exponents

Vocabulary
base • exponent • variable • squared • cubed • to the (exponent)-th power

1. The _____ tells you how many times the base is used as a factor.

2. If the base has an exponent of 3, we say the base is _____.

3. The _____ tells you what number is being multiplied in exponent form.

4. If we do not know the value of a number, we use a letter, called a(n) _____, to represent the unknown number.

Example	Student Practice
1. Write in exponent form.	**2.** Write in exponent form.
(a) $9(9)(9)$	**(a)** $-3(-3)(-3)(-3)$
The base is 9 and it is used as a factor three times. So, the exponent is 3.	
$9(9)(9) = 9^3$	
	(b) $-6(-6)(-6)(-6)(-6)(-6)(-6)(-6)$
(b) $-7(-7)(-7)(-7)(-7)$	
The base is -7 and it is used as a factor five times. The answer must contain parentheses.	
$-7(-7)(-7)(-7)(-7) = (-7)^5$	**(c)** $(n)(n)(n)(n)(n)(n)$
(c) $(y)(y)(y)$	
$(y)(y)(y) = y^3$	

Vocabulary Answers: 1. exponent 2. cubed 3. base 4. variable

Example	Student Practice
3. Evaluate.	**4.** Evaluate.
(a) 2^5	**(a)** 4^3
$2^5 = (2)(2)(2)(2)(2) = 32$	
(b) $2^3 + 4^4$	
First evaluate each power. Then add.	**(b)** $1^7 + 6^2$
$2^3 + 4^4 = 8 + 256$	
$\qquad\qquad = 264$	

5. Evaluate.	**6.** Evaluate.
(a) $(-2)^3$	**(a)** $(-4)^3$
The answer is negative since the base is negative and the exponent is odd, $(-2)^3 = -8$.	
(b) $(-4)^6$	**(b)** $(-4)^4$
The answer is positive since the exponent 6 is even, $(-4)^6 = +4096$.	
(c) -3^6	**(c)** -4^4
The negative sign is not contained within parentheses. Thus, find 3^6 and take the opposite of that value, $-3^6 = -729$.	
(d) $-(5^4)$	**(d)** $-(4^4)$
The negative sign is outside the parentheses, $-(5^4) = -625$.	

Example	Student Practice
7. Evaluate.	**8.** Evaluate.

(a) $\left(\dfrac{1}{2}\right)^4$

$$\left(\dfrac{1}{2}\right)^4 = \left(\dfrac{1}{2}\right)\left(\dfrac{1}{2}\right)\left(\dfrac{1}{2}\right)\left(\dfrac{1}{2}\right)$$
$$= \dfrac{1}{16}$$

(b) $(0.2)^4$

$$(0.2)^4 = (0.2)(0.2)(0.2)(0.2)$$
$$= 0.0016$$

(c) $\left(\dfrac{2}{5}\right)^3$

$$\left(\dfrac{2}{5}\right)^3 = \left(\dfrac{2}{5}\right)\left(\dfrac{2}{5}\right)\left(\dfrac{2}{5}\right)$$
$$= \dfrac{8}{125}$$

(d) $(3)^3 (2)^5$

First we evaluate each power, then we multiply.

$$(3)^3 (2)^5 = (27)(32)$$
$$= 864$$

(e) $2^3 - 3^4$

$$2^3 - 3^4 = 8 - 81$$
$$= -73$$

Student Practice column:

(a) $\left(\dfrac{1}{5}\right)^3$

(b) $(0.3)^3$

(c) $\left(\dfrac{4}{7}\right)^3$

(d) $(2)^4 (5)^2$

(e) $10^3 - 2^{10}$

Extra Practice

1. Write the product in exponent form. Do not evaluate.

    $(-ab)(-ab)$

2. Evaluate.

    $\left(-\dfrac{1}{2}\right)^3$

3. Evaluate.

    $-6^2 - (-2)^2$

4. Evaluate.

    $7^2 - (-2)^3$

Concept Check

Explain the difference between $(-2)^6$ and -2^6. How do you decide if the answers are positive or negative?

Chapter 1 Real Numbers and Variables
1.5 The Order of Operations

Vocabulary
order of operations   •   parentheses   •   multiply and divide   •   add and subtract

1. When simplifying an expression, the last priority is to _____ numbers from left to right.

2. When simplifying an expression, the first priority is to do all operations inside _____.

3. The list of operations to do first is called the _____.

4. When simplifying an expression, the third priority is to _____ numbers from left to right.

Example	Student Practice
1. Evaluate. $8 \div 2 \cdot 3 + 4^2$	**2.** Evaluate. $36 \div 3 \cdot 2 + 6^2$

Evaluate $4^2 = 16$ because the highest priority in this problem is raising to a power.

$$8 \div 2 \cdot 3 + 4^2 = 8 \div 2 \cdot 3 + 16$$

Next, multiply and divide from left to right.

$$8 \div 2 \cdot 3 + 16 = 4 \cdot 3 + 16$$
$$= 12 + 16$$

Finally, add.

$$12 + 16 = 28$$

Vocabulary Answers: 1. add and subtract 2. parentheses 3. order of operations 4. multiply and divide

Example	Student Practice
3. Evaluate. $(-3)^3 - 2^4$	**4.** Evaluate. $(-5)^2 - 1^{10}$

3. Evaluate. $(-3)^3 - 2^4$

The highest priority is to raise the expressions to the appropriate powers.

In $(-3)^3$ we are cubing the number -3 to obtain -27.

$$(-3)^3 - 2^4 = -27 - 2^4$$

Be careful; -2^4 is not $(-2)^4$. We raise 2 to the fourth power. Then, add and subtract from left to right.

$$(-3)^3 - 2^4 = -27 - 16$$
$$= -43$$

4. Evaluate. $(-5)^2 - 1^{10}$

---

5. Evaluate. $2 \cdot (2-3)^3 + 6 \div 3 + (8-5)^2$

Combine the numbers inside the parentheses.

$$2 \cdot (2-3)^3 + 6 \div 3 + (8-5)^2$$
$$= 2 \cdot (-1)^3 + 6 \div 3 + 3^2$$

We need parentheses for -1 because of the negative sign, but they are not needed for 3. Next, raise to a power.

$$2 \cdot (-1)^3 + 6 \div 3 + 3^2 = 2 \cdot (-1) + 6 \div 3 + 9$$

Next, multiply and divide from left to right, $2 \cdot (-1) + 6 \div 3 + 9 = -2 + 2 + 9$.

Finally, add and subtract from left to right, $-2 + 2 + 9 = 9$.

6. Evaluate. $3 \cdot (3-5)^2 + 15 \div 5 + (7-5)^3$

Example	Student Practice

7. Evaluate. $\left(-\dfrac{1}{5}\right)\left(\dfrac{1}{2}\right)-\left(\dfrac{3}{2}\right)^2$

The highest priority is to raise $\dfrac{3}{2}$ to the second power, $\left(\dfrac{3}{2}\right)^2 = \left(\dfrac{3}{2}\right)\left(\dfrac{3}{2}\right) = \dfrac{9}{4}$.

$$\left(-\dfrac{1}{5}\right)\left(\dfrac{1}{2}\right)-\left(\dfrac{3}{2}\right)^2 = \left(-\dfrac{1}{5}\right)\left(\dfrac{1}{2}\right)-\dfrac{9}{4}$$

Next we multiply.

$$\left(-\dfrac{1}{5}\right)\left(\dfrac{1}{2}\right)-\dfrac{9}{4} = -\dfrac{1}{10}-\dfrac{9}{4}$$

We need to write each fraction as an equivalent fraction with the LCD of 20.

$$\left(-\dfrac{1}{5}\right)\left(\dfrac{1}{2}\right)-\dfrac{9}{4} = -\dfrac{1\cdot 2}{10\cdot 2}-\dfrac{9\cdot 5}{4\cdot 5}$$

$$= -\dfrac{2}{20}-\dfrac{45}{20}$$

Finally, subtract.

$$-\dfrac{2}{20}-\dfrac{45}{20} = -\dfrac{47}{20} \text{ or } -2\dfrac{7}{20}$$

8. Evaluate. $\left(\dfrac{3}{4}\right)^2 + \dfrac{14}{15}\left(-\dfrac{5}{7}\right)$

Extra Practice

1. Evaluate. Simplify all fractions.

$$1 + 2 - 3 \times 4 \div 6$$

2. Evaluate. Simplify all fractions.

$$50 \div 5 \times 2 + (11 - 8)^2$$

3. Evaluate. Simplify all fractions.

$$\left(\frac{1}{2}\right)^2 \div \left(\frac{1}{2}\right)^3$$

4. Evaluate. Simplify all fractions.

$$\left(\frac{2}{3}\right)^2 (-18) + \frac{3}{7} \div \frac{5}{21}$$

Concept Check

Explain in what order you would perform the calculations to evaluate the expression $4 - (-3 + 4)^3 + 12 \div (-3)$.

Name: _____   Date: _____

Instructor: _____   Section: _____

Chapter 1 Real Numbers and Variables
1.6 Using the Distributive Property to Simplify Algebraic Expressions

Vocabulary
algebraic expression • term • distributive property • factors

1. The _____ states that for all real numbers a, b, and c, $a(b+c) = ab + ac$.

2. A(n) _____ is a quantity that contains numbers and variables.

3. Two or more algebraic expressions that are multiplied are called _____.

4. A(n) _____ is a number, variable, or a product of numbers and variables.

Example	Student Practice
1. Multiply.	**2.** Multiply.
(a) $5(a+b)$	**(a)** $6(x+3y)$
Multiply the factor $(a+b)$ by the factor 5 using the distributive property.	
$5(a+b) = 5a + 5b$	**(b)** $-4(2m+5n)$
(b) $-3(3x+2y)$	
$-3(3x+2y) = -3(3x) + (-3)(2y)$ $= -9x - 6y$	
3. Multiply. $-(a-2b)$	**4.** Multiply. $-(m-6n)$
$-(a-2b) = (-1)(a-2b)$ $= (-1)(a) + (-1)(-2b)$ $= -a + 2b$	

Vocabulary Answers: 1. distributive property 2. algebraic expression 3. factor 4. term

Example	Student Practice
5. Multiply.	**6.** Multiply.

(a) $\dfrac{2}{3}\left(x^2 - 6x + 8\right)$ **(a)** $\dfrac{3}{4}\left(m^2 - 8m + 5\right)$

Apply the distributive property.

$$\dfrac{2}{3}\left(x^2 - 6x + 8\right)$$

$$= \left(\dfrac{2}{3}\right)\left(1x^2\right) + \left(\dfrac{2}{3}\right)(-6x) + \left(\dfrac{2}{3}\right)(8)$$

$$= \dfrac{2}{3}x^2 + (-4x) + \dfrac{16}{3}$$

(b) $1.3\left(x^2 + 1.3x + 0.7\right)$

$$= \dfrac{2}{3}x^2 - 4x + \dfrac{16}{3}$$

(b) $1.4\left(a^2 + 2.5a + 1.8\right)$

$$1.4\left(a^2 + 2.5a + 1.8\right)$$

$$= 1.4\left(1a^2\right) + (1.4)(2.5a) + (1.4)(1.8)$$

$$= 1.4a^2 + 3.5a + 2.52$$

---

7. Multiply. $-2x(3x + y - 4)$ **8.** Multiply. $-5x(x + 3y - 6)$

$$-2x(3x + y - 4)$$

$$= -2(x)(3)(x) - 2(x)(y) - 2(x)(-4)$$

$$= -2(3)(x)(x) - 2(xy) - 2(-4)(x)$$

$$= -6x^2 - 2xy + 8x$$

---

9. Multiply. $\left(2x^2 - x\right)(-3)$ **10.** Multiply. $\left(4y - 3y^2\right)(-2)$

$$\left(2x^2 - x\right)(-3) = 2x^2(-3) + (-x)(-3)$$

$$= -6x^2 + 3x$$

Example	Student Practice

11. A farmer has a rectangular field that is 300 feet wide. One portion of the field is $2x$ feet long. The other portion of the field is $3y$ long. Use the distributive property to find an expression for the area of this field.

First we draw a picture of a field that is 300 feet wide and $2x + 3y$ feet long.

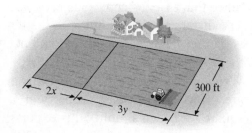

To find the area of the field, we multiply the width times the length.

$$300(2x + 3y)$$

Then, apply the distributive property.

$$300(2x + 3y) = 300(2x) + 300(3y)$$

Finally, simplify.

$$300(2x) + 300(3y) = 600x + 900y$$

Thus the area of the field in square feet is $600x + 900y$.

12. A convenience store is 500 feet wide. One portion of the store is $7x$ feet long. The other portion of the store is $8y$ feet long. Use the distributive property to find an expression for the area of this store.

Extra Practice

1. Multiply. Use the distributive property.

$$-2(6x+3)$$

2. Multiply. Use the distributive property.

$$\frac{1}{2}\left(\frac{1}{3}x+\frac{2}{5}y\right)$$

3. Multiply. Use the distributive property.

$$(6x-4y+2z)(-3)$$

4. Mr. Jorgensen's backyard is a rectangle that is 30 feet wide. He recently built a fence that extends across his backyard. The distance from his house to the fence is 15 feet. The distance from the fence to the back edge of his property is $7n$. Write an expression that represents the area of Mr. Jorgensen's property in square feet, then simplify this expression.

Concept Check

Multiply. Use the distributive property and explain how you would multiply to obtain the answer for $\left(-\frac{3}{7}\right)\left(21x^2-14x+3\right)$.

Chapter 1 Real Numbers and Variables
1.7 Combining Like Terms

Vocabulary
combine like terms   •   like terms   •   combining   •   term

1. A _____ is a number, a variable, or a product of numbers and variables separated by plus or minus signs in an algebraic expression.

2. We can add or subtract quantities that are like quantities, which is called _____ like quantities.

3. To _____ is to add or subtract like terms in an algebraic expression.

4. _____ are terms that have identical variables and exponents.

Example	Student Practice
1. List the like terms of each expression.	**2.** List the like terms of each expression.
(a) $5x - 2y + 6x$	(a) $4x - 6y + 2x + 7$
$5x$ and $6x$ are like terms.	
(b) $2x^2 - 3x - 5x^2 - 8x$	(b) $4x^2 + 5x - 6x^2 + 9x$
$2x^2$ and $-5x^2$ are like terms. $-3x$ and $-8x$ are like terms. Note that x^2 and x are not like terms.	
3. Combine like terms.	**4.** Combine like terms.
(a) $-4x^2 + 8x^2$	(a) $10z^5 - 3z^5$
Each term contains the factor x^2.	
$-4x^2 + 8x^2 = (-4 + 8)x^2 = 4x^2$	
(b) $5x + 3x + 2x$	(b) $5z - 7z + 4z$
$5x + 3x + 2x = (5 + 3 + 2)x = 10x$	

Vocabulary Answers: 1. term 2. combining 3. combine like terms 4. like terms

Example	Student Practice
5. Simplify. $5a^2 - 2a^2 + 6a^2$	**6.** Simplify. $x^3 - 7x^3 + 9x^3$

5. Simplify. $5a^2 - 2a^2 + 6a^2$

$$5a^2 - 2a^2 + 6a^2 = (5 - 2 + 6)a^2$$
$$= 9a^2$$

7. Simplify.

   (a) $5.6a + 2b + 7.3a - 6b$

   We combine the a terms and the b terms separately.

   $5.6a + 2b + 7.3a - 6b = 12.9a - 4b$

   (b) $3x^2y - 2xy^2 + 6x^2y$

   Note that x^2y and xy^2 are not like terms because of different powers.

   $3x^2y - 2xy^2 + 6x^2y = 9x^2y - 2xy^2$

   (c) $2a^2b + 3ab^2 - 6a^2b^2 - 8ab$

   These terms cannot be combined; there are no like terms in this expression.

   $2a^2b + 3ab^2 - 6a^2b^2 - 8ab$

8. Simplify.

   (a) $4.1m + 6n + 3.5m + 1.2n$

   (b) $6a^2b - 4ab^3 + 2a^2b$

   (c) $6xy + 2x^2y - 9xy^2 + 5x^2y^2$

9. Simplify. $3a - 2b + 5a^2 + 6a - 8b - 12a^2$

There are three pairs of like terms. You can rearrange the terms so that like terms are together.

$$\underbrace{3a + 6a}_{a\text{ terms}} \underbrace{- 2b - 8b}_{b\text{ terms}} + \underbrace{5a^2 - 12a^2}_{a^2\text{ terms}}$$
$$= 9a - 10b - 7a^2$$

10. Simplify. $9x - 9c + 2t + 4c - 4t + 5x$

Example	Student Practice
11. Simplify. $\dfrac{3}{4}x^2 - 5y - \dfrac{1}{8}x^2 + \dfrac{1}{3}y$	**12.** Simplify. $\dfrac{2}{5}x - \dfrac{6}{7}y + 3x + \dfrac{1}{2}y$

We need the least common denominator for the x^2 terms, which is 8.

$$\frac{3}{4}x^2 - \frac{1}{8}x^2 = \frac{3\cdot 2}{4\cdot 2}x^2 - \frac{1}{8}x^2$$
$$= \frac{6}{8}x^2 - \frac{1}{8}x^2$$
$$= \frac{5}{8}x^2$$

The least common denominator for the y terms is 3.

$$-\frac{5}{1}y + \frac{1}{3}y = \frac{-5\cdot 3}{1\cdot 3}y + \frac{1}{3}y$$
$$= \frac{-15}{3}y + \frac{1}{3}y$$
$$= -\frac{14}{3}y$$

Thus, our solution is $\dfrac{5}{8}x^2 - \dfrac{14}{3}y$.

13. Simplify. $6(2x+3xy)-8x(3-4y)$	**14.** Simplify. $5(2c+7cd)-3c(9-4d)$

First remove the parentheses using the distributive property. Then combine like terms.

$$6(2x+3xy)-8x(3-4y)$$
$$= 12x + 18xy - 24 + 32xy$$
$$= -12x + 50xy$$

Extra Practice

1. Combine like terms.

$$9a + 10 - 2a^2 - 11a + 2 + 6a^2$$

2. Combine like terms.

$$\frac{3}{7}x^2 - \frac{1}{2}y + \frac{3}{14}x^2 - \frac{1}{8}y$$

3. Simplify. Use the distributive property to remove parentheses; then combine like terms.

$$-2(7a - 3b) + 4(-2a + 4b)$$

4. Simplify. Use the distributive property to remove parentheses; then combine like terms.

$$2a(b - 4c) - 3c(-7a + b - 10d)$$

Concept Check

Explain how you would remove parentheses and then combine like terms to obtain the answer for $1.2(3.5x - 2.2y) - 4.5(2.0x + 1.5y)$.

Chapter 1 Real Numbers and Variables
1.8 Using Substitution to Evaluate Algebraic Expressions and Formulas

Vocabulary
evaluate • substituted • perimeter • area • right angles • altitude • rectangle
parallelogram • square • trapezoid • triangle • circle • circumference

1. A(n) _____ is a rectangle with all four sides equal.

2. _____ is the distance around a circle.

3. _____ is a measure of the amount of surface in a region.

4. You will use the order of operations to _____ variable expressions.

Example	Student Practice
1. Evaluate $\frac{2}{3}x - 5$ for $x = -6$.	**2.** Evaluate $\frac{3}{4}x - 7$ for $x = -24$.
Substitute -6 for x. $$\frac{2}{3}x - 5 = \frac{2}{3}(-6) - 5$$ $$= -4 - 5$$ $$= -9$$	

3. Evaluate for $x = -3$.
 	
(a) $2x^2$
    Here the value x is squared.
    $$2x^2 = 2(-3)^2$$
    $$= 2(9)$$
    $$= 18$$
(b) $(2x)^2$
    Here the value $(2x)$ is squared.
    $$(2x)^2 = \left[2(-3)\right]^2$$
    $$= (-6)^2$$
    $$= 36$$
	4. Evaluate for $x = -5$.
	
	(a) $3y^2$
	
	(b) $(3y)^2$

Vocabulary Answers: 1. square 2. circumference 3. area 4. evaluate

Example	Student Practice
5. Evaluate $x^2 + 3x$ for $x = -4$.	**6.** Evaluate $5x^2 + 4x$ for $x = -3$.

Replace each x by -4.

$$x^2 + 3x = (-4)^2 + 3(-4)$$
$$= 16 + 3(-4)$$
$$= 16 - 12$$
$$= 4$$

---

7. Evaluate $x^3 + 2xy - 3x + 1$ for $x = 2$ and $y = -\dfrac{1}{4}$.

Replace x with 2 and y with $-\dfrac{1}{4}$.

$$x^3 + 2xy - 3x + 1$$
$$= (2)^3 + 2(2)\left(-\frac{1}{4}\right) - 3(2) + 1$$
$$= 8 + (-1) - 6 + 1$$
$$= 8 + (-1) + (-6) + 1$$
$$= 9 + (-7)$$
$$= 2$$

8. Evaluate $7x^2 - 10xy - 50y^2$ for $x = -4$ and $y = \dfrac{1}{5}$.

---

9. Find the area of a triangle with a base of 16 centimeters (cm) and a height of 12 centimeters (12).

Substitute $a = 12$ cm and $b = 16$ cm in $A = \dfrac{1}{2}ab$.

$$A = \frac{1}{2}(12 \text{ cm})(16 \text{ cm})$$
$$= (6)(16)(\text{cm})^2$$
$$= 96 \text{ square centimeters}$$

The area of the triangle is 96 square centimeters or 96 cm^2.

10. Find the area of a triangle with a height of 9 yards and a base of 3 yards.

Example	Student Practice
11. Find the area of a circle if the radius is 2 inches.	**12.** Find the area of a circle if the radius is 4 yards.

Write the formula and substitute the given values for the letters.

$$A = \pi r^2 \approx (3.14)(2 \text{ inches})^2$$

Raise to a power. Then multiply.

$$(3.14)(2 \text{ inches})^2 = (3.14)(4)(\text{in.})^2$$
$$= 12.56 \text{ in.}^2$$

Thus the area is 12.56 square inches or 12.56 in.^{2}

---

13. What is the Celsius temperature when the Fahrenheit temperature is $F = -22°$?	**14.** What is the Celsius temperature when the Fahrenheit temperature is $F = -40°$? Use the formula $C = \dfrac{5}{9}(F - 32)$.

Use the formula. Substitute -22 for F in the formula.

$$C = \frac{5}{9}(F - 32)$$
$$= \frac{5}{9}\left[(-22) - 32\right]$$

Combine the numbers inside the brackets. Then simplify and multiply.

$$\frac{5}{9}\left[(-22) - 32\right] = \frac{5}{9}(-54)$$
$$= (5)(-6)$$
$$= -30$$

The temperature is $-30°$ Celsius or $-30°\text{C}$.

Example	Student Practice
15. You are driving on a highway in Mexico. It has a posted maximum speed of 100 kilometers per hour. You are driving at 61 miles per hour. Are you exceeding the speed limit?	**16.** You are driving on a Canadian highway. The posted maximum speed limit is 110 kilometers per hour. You are driving at 70 miles per hour. Are you exceeding the speed limit?

Use the formula. Replace r by 61. Then multiply the numbers.

$$k \approx 1.61r$$
$$= (1.61)(61)$$
$$= 98.21$$

You are driving approximately 98 kilometers per hour. You are not exceeding the speed limit.

Extra Practice

1. Evaluate $\dfrac{2}{7}y - 7$ for $y = 14$.

2. Evaluate $2x^2 + 3x - 8$ for $x = 3$.

3. Evaluate $\dfrac{2a^2 - 3b}{b^2}$ for $a = -1$ and $b = 2$.

4. A circular patch of grass has a radius of 6 meters. What is the area of the patch of grass, rounded to the nearest square meter?

Concept Check
Explain how you would find the area of a circle if you know its diameter is 12 meters.

Chapter 1 Real Numbers and Variables
1.9 Grouping Symbols

Vocabulary
grouping symbols   •   fraction bars   •   distributive property   •   like terms

1. _____ are also considered grouping symbols.

2. Use the _____ and the rules for real numbers to remove grouping symbols

3. Many expressions in algebra use _____ such as parentheses, brackets, and braces.

Example	Student Practice
1. Simplify. $3\left[6-2(x+y)\right]$	**2.** Simplify. $4\left[6x-5(x+3)\right]$
We want to remove the innermost parentheses first. Therefore, we first use the distributive property to simplify.	
$3\left[6-2(x+y)\right]=3\left[6-2x-2y\right]$	
Use the distributive property again.	
$3\left[6-2x-2y\right]=18-6x-6y$	
3. Simplify. $-2\left[3a-(b+2c)+(d-3e)\right]$	**4.** Simplify. $-4\left[5x-(2y-7)+(3x-9)\right]$
Remove both inner sets of parentheses.	
$-2\left[3a-(b+2c)+(d-3e)\right]$ $=-2\left[3a-b-2c+d-3e\right]$	
Now remove the brackets by multiplying each term by -2.	
$-2\left[3a-b-2c+d-3e\right]$ $=-6a+2b+4c-2d+6e$	

Vocabulary Answers: 1. fraction bars 2. distributive property 3. grouping symbols

Example	Student Practice
5. Simplify. $$2\left[3x-(y+w)\right]-3\left[2x+2(3y-2w)\right]$$ $$= 2\left[3x-y-w\right]-3\left[2x+6y-4w\right]$$ $$= 6x-2y-2w-6x-18y+12w$$ $$= -20y+10w \text{ or } 10w-20y$$	**6.** Simplify. $$5\left[x-3(4-2x)\right]+(-3)\left[7x-6(x+1)\right]$$

7. Simplify. $-3\left\{7x-2\left[x-(2x-1)\right]\right\}$

Remember to remove the innermost grouping symbols first.

$$-3\left\{7x-2\left[x-(2x-1)\right]\right\}$$
$$= -3\left\{7x-2\left[x-2x+1\right]\right\}$$
$$= -3\left\{7x-2\left[-x+1\right]\right\}$$
$$= -3\left\{7x+2x-2\right\}$$
$$= -3\left\{9x-2\right\}$$
$$= -27x+6$$

8. Simplify. $-4\left\{6x-\left[3x-(2-x)\right]\right\}$

Extra Practice

1. Simplify. Remove grouping symbols and combine like terms. $-2a-3(a+3b)$

2. Simplify. Remove grouping symbols and combine like terms.
$$-2\left[3(x-y)-2(x+y)\right]$$

3. Simplify. Remove grouping symbols and combine like terms.
$$2b-\left\{3a+4\left[2a-(-2b+3a)\right]\right\}$$

4. Simplify. Remove grouping symbols and combine like terms.
$$-2\left\{5x^2+2\left[3x-(5-x)\right]\right\}$$

Concept Check

Explain how you would simplify the following expression to combine like terms whenever possible. $3\left\{2-3\left[4x-2(x+3)+5x\right]\right\}$

MATH COACH

Mastering the skills you need to do well on the test.

Watch the MATH COACH videos in MyMathLab® or on YouTube™ while you work the problems below. These helpful hints will help you avoid making common errors on test problems.

Evaluating Exponential Expressions When the Base Is a Fraction—Problem 9   Simplify $\left(\dfrac{2}{3}\right)^4$.

> **Helpful Hint:** Always write out the repeated multiplication.
> - If a number is raised to the fourth power, then we write out the multiplication with that number appearing as a factor a total of four times.
> - When the base is a fraction, this means that we must multiply the numerators and then multiply the denominators.

Did you rewrite the problem as $\left(\dfrac{2}{3}\right)\left(\dfrac{2}{3}\right)\left(\dfrac{2}{3}\right)\left(\dfrac{2}{3}\right)$?

Yes ____   No ____

If you answered No to this question, please complete this step now.

Did you multiply $2\times2\times2\times2$ in the numerator and $3\times3\times3\times3$ in the denominator? Yes ____   No ____

If you answered No, make this correction and complete the calculations.

If you answered Problem 9 incorrectly, go back and rework the problem using these suggestions.

Using the Order of Operations with Numerical Expressions—Problem 11

Simplify $3(4-6)^3+12\div(-4)+2$.

> **Helpful Hint:** First, do all operations inside parentheses. Second, raise numbers to a power. Then multiply and divide numbers from left to right. As the last step, add and subtract numbers from left to right.

Did you first combine $4-6$ to obtain -2?
Yes ____   No ____

If you answered No, perform the operation inside the parentheses first.

Next, did you calculate $(-2)^3=-8$ before multiplying by 3?
Yes ____   No ____

If you answered No, go back and evaluate the exponential expression.

Did you multiply and divide before adding?
Yes ____   No ____

If you answered No, remember that multiplication and division must be performed before addition and subtraction once the operations inside the parentheses are done and exponents are evaluated.

Evaluate Algebraic Expressions for a Specified Value—Problem 19

Evaluate $3x^2 - 7x - 11$ for $x = -3$.

> **Helpful Hint:** When you replace a variable by a particular value, place parentheses around that value. Then use the order of operations to evaluate the expression.

Did you rewrite the expression as $3(-3)^2 - 7(-3) - 11$, using parentheses to complete the substitution?

Yes _____   No _____

If you answered No, please go back and perform that substitution step using parentheses around the specified value.

Did you next raise -3 to the second power to obtain $3(9) - 7(-3) - 11$?

Yes _____   No _____

If you answered No, review the order of operations and complete this step.

Remember that multiplication must be performed before addition and subtraction.

If you answered Problem 19 incorrectly, go back and rework this problem using these suggestions.

Simplifying Algebraic Expressions with Many Grouping Symbols—Problem 26

Simplify $-3\{a + b[3a - b(1-a)]\}$.

> **Helpful Hint:** Work from the inside out. Remove the innermost symbols, ( ), first. Then remove the next level of innermost symbols, [ ]. Finally remove the outermost symbols, { }. Be careful to avoid sign errors.

Did you first obtain the expression $-3\{a + b[3a - b + ab]\}$ when you removed the innermost parentheses?

Yes _____   No _____

If you answered No, go back to the original problem and use the distributive property to remove the innermost grouping symbol, the parentheses.

Did you next obtain the expression $-3\{a + 3ab - b^2 + ab^2\}$ when you removed the brackets?

Yes _____   No _____

If you answered No, use the distributive property to distribute b on the outside of the bracket, [ ], to each term inside the bracket.

Finally, use the distributive property in the last step to distribute the -3 across all of the terms inside the outermost brackets, { }. Be careful with the $+/-$ signs.

Now go back and rework the problem using these suggestions.

Chapter 2 Equations and Inequalities
2.1 The Addition Principle of Equality

Vocabulary
Equation • solution • equivalent equations • solving the equation • identity
satisfies • checking • addition principle • opposite in sign • additive inverse

1. A number that is opposite in sign to another number is called its _____.

2. A(n) _____ is a statement in which the equals sign $(=)$ is used to indicate that two expressions are equal.

3. If the same value appears on both sides of the equals sign, the equation is called a(n) _____.

4. If the same number is added to both sides of an equation, the _____ states that the results on both sides are equal.

Example	Student Practice
1. Solve for x. $x+16 = 20$	**2.** Solve for x. $x+8 = 42$

Use the addition principle to add -16 to both sides and simplify.

$$x+16 = 20$$
$$x+16+(-16) = 20+(-16)$$
$$x+0 = 4$$
$$x = 4$$

Substitute the value found for x into the original equation to verify the solution.

$$x+16 = 20$$
$$4+16 \overset{?}{=} 20$$
$$20 = 20$$

The solution checks.

Vocabulary Answers: 1. additive inverse 2. equation 3. identity 4. addition principle

Example	Student Practice
3. Solve for x. $1.5 + 0.2 = 0.3 + x + 0.2$	**4.** Solve for x. $2.7 - 0.2 = 0.4 + x - 0.6$

3. Solve for x. $1.5 + 0.2 = 0.3 + x + 0.2$

Simplify by adding.

$1.5 + 0.2 = 0.3 + x + 0.2$
$1.7 = x + 0.5$

Add -0.5 to both sides.

$1.7 + (-0.5) = x + 0.5 + (-0.5)$

Simplify.

$1.2 = x$

The check is left to the student.

---

5. Is 10 the solution to the equation
$-15 + 2 = x - 3$? If not, find the solution.

Substitute 10 for x in the equation.

$-15 + 2 = x - 3$

$-15 + 2 \overset{?}{=} 10 - 3$

$13 \neq 7$

The values are not equal. Thus, 10 is not the solution. Solve the original equation to find the solution. Start by simplifying.

$-15 + 2 = x - 3$
$-13 = x - 3$

Add 3 to both sides.

$-13 = x - 3$
$-13 + 3 = x - 3 + 3$
$-10 = x$

The check is left to the student.

6. Is 26 the solution to the equation
$x - 12 = 28 - 4$? If not, find the solution.

Example	Student Practice
7. Find the value of x that satisfies the equation. $$\frac{1}{5} + x = -\frac{1}{10} + \frac{1}{2}$$	**8.** Find the value of x that satisfies the equation. $$\frac{1}{3} + x = -\frac{1}{21} + \frac{1}{7}$$

To be combined, the fractions must have common denominators. Rewrite each fraction as an equivalent fraction with a denominator of 10 and simplify.

$$\frac{1}{5} \cdot \frac{2}{2} + x = -\frac{1}{10} + \frac{1}{2} \cdot \frac{5}{5}$$

$$\frac{2}{10} + x = -\frac{1}{10} + \frac{5}{10}$$

$$\frac{2}{10} + x = \frac{4}{10}$$

Add $-\dfrac{2}{10}$ to each side.

$$\frac{2}{10} + \left(-\frac{2}{10}\right) + x = \frac{4}{10} + \left(-\frac{2}{10}\right)$$

$$x = \frac{2}{10} = \frac{1}{5}$$

Check.

$$\frac{1}{5} + x = -\frac{1}{10} + \frac{1}{2}$$

$$\frac{1}{5} + \left(\frac{1}{5}\right) \overset{?}{=} -\frac{1}{10} + \frac{1}{2}$$

$$\frac{2}{5} \overset{?}{=} -\frac{1}{10} + \frac{5}{10}$$

$$\frac{2}{5} \overset{?}{=} \frac{4}{10}$$

$$\frac{2}{5} = \frac{2}{5}$$

The solution checks.

Extra Practice

1. Solve for x. Check your answers.

   $$14 = 8 + x$$

2. Solve for x. Check your answers.

   $$1.7 + 4.2 + x = 19.23 - 9.8$$

3. Is 6 the solution to the equation $x + 7 = 14 - 19 + 6$? If it is not, find the solution.

4. Is 28 the solution to the equation $-17 + x - 4 = 12 - 8 + 3$? If it is not, find the solution.

Concept Check

Explain how you would check to verify whether $x = 3.8$ is the solution to $-1.3 + 1.6 + 3x = -6.7 + 4x + 3.2$.

Chapter 2 Equations and Inequalities
2.2 The Multiplication Principle of Equality

Vocabulary
terminating decimal • multiplicative inverse • multiplication principle • division principle

1. If both sides of an equation are multiplied by the same nonzero number, the _____ states that the results on both sides are equal.

2. The _____ states that dividing both sides of an equation by the same nonzero number results in an equivalent equation.

3. For any nonzero number a, the _____ is $\dfrac{1}{a}$.

4. A _____ is a decimal with a definite number of digits.

Example	Student Practice
1. Solve for x. $\dfrac{1}{3}x = -15$	**2.** Solve for x. $-\dfrac{1}{7}x = -3$

We know that $(3)\left(\dfrac{1}{3}\right) = 1$. Multiply each side of the equation by 3 to isolate x.

$$3\left(\dfrac{1}{3}x\right) = 3(-15)$$

$$\left(\dfrac{3}{1}\right)\left(\dfrac{1}{3}\right)x = -45$$

$$x = -45$$

Check.

$$\dfrac{1}{3}(-45) \overset{?}{=} -15$$

$$-15 = -15$$

The solution checks.

Vocabulary Answers: 1. multiplication principle 2. division principle 3. multiplicative inverse
4. terminating decimal

Example	Student Practice
3. Solve for x. $5x = 125$	**4.** Solve for x. $6x = 42$

3. Solve for x. $5x = 125$

Divide both sides by 5.

$$\frac{5x}{5} = \frac{125}{5}$$
$$x = 25$$

Check.

$$5x = 125$$
$$5(25) \overset{?}{=} 125$$
$$125 = 125$$

The solution checks.

---

4. Solve for x. $6x = 42$

---

5. Solve for x. $9x = 60$

Divide both sides by 9 and simplify.

$$9x = 60$$
$$\frac{9x}{9} = \frac{60}{9}$$
$$x = \frac{20}{3}$$

The check is left to the student.

---

6. Solve for x. $16x = 30$

---

7. Solve for x. $-3x = 48$

Divide both sides by -3.

$$\frac{-3x}{-3} = \frac{48}{-3}$$
$$x = -16$$

The check is left to the student.

---

8. Solve for x. $-9x = 63$

Example	Student Practice
9. Solve for x. $-x = -24$	**10.** Solve for x. $-x = 15$

9. Solve for x. $-x = -24$

Rewrite the equation. Note that $-1x$ is the same as $-x$.

$-1x = -24$

Divide both sides by the coefficient -1.

$$\frac{-1x}{-1} = \frac{-24}{-1}$$
$$x = 24$$

The check is left to the student.

---

10. Solve for x. $-x = 15$

---

11. Solve for x. $-78 = 5x - 8x$

Combine the like terms on the right side.

$-78 = 5x - 8x$
$-78 = -3x$

Divide both sides by the coefficient -3.

$$\frac{-78}{-3} = \frac{-3x}{-3}$$
$$26 = x$$

The check is left to the student.

---

12. Solve for x. $24 = 4x - 8x$

---

13. Solve for x. $31.2 = 6.0x - 0.8x$

Combine the like terms on the right side.

$31.2 = 6.0x - 0.8x$
$31.2 = 5.2x$

Divide both sides by 5.2.

$$\frac{31.2}{5.2} = \frac{5.2x}{5.2}$$
$$6 = x$$

---

14. Solve for x. $15.4 = 4.8x - 2.6x$

Extra Practice

1. Solve for x. Check your answers.

   $-13 = -x$

2. Solve for x. Check your answers.

   $-5 = \dfrac{x}{5}$

3. Is 7 the solution to the equation $\dfrac{3}{7}x = 3$? If it is not, find the correct solution.

4. Is 0.12 the solution to the equation $3x = -0.36$? If it is not, find the correct solution.

Concept Check

Explain how you would check to verify whether $x = 36\dfrac{2}{3}$ is the solution to $-22 = -\dfrac{3}{5}x$.

Name: _____   Date: _____

Instructor: _____   Section: _____

Chapter 2 Equations and Inequalities
2.3 Using the Addition and Multiplication Principles Together

Vocabulary
distributive property   •   like terms   •   multiplication principle   •   addition principle

1. To solve an equation of the form $ax^2 + bx = c$, we must use both the addition principle and the _____.

2. If the variable appears on both sides of the equation, apply the _____ to the variable term to collect the variable terms on one side of the equation.

3. If an equation contains parentheses, first apply the _____ to remove the parentheses.

4. Where it is possible, first collect _____ on one or both sides of the equation.

Example	Student Practice
1. Solve for x. $5x + 3 = 18$	**2.** Solve for x. $6x - 8 = 4$

Use the addition principle to add -3 to both sides of the equation and simplify.

$$5x + 3 = 18$$
$$5x + 3 + (-3) = 18 + (-3)$$
$$5x = 15$$

Use the division principle to divide both sides by 5 and simplify.

$$5x = 15$$
$$\frac{5x}{5} = \frac{15}{5}$$
$$x = 3$$

Check. $5(3) + 3 \overset{?}{=} 18$

$$15 + 3 \overset{?}{=} 18$$
$$18 = 18$$

Vocabulary Answers: 1. multiplication principle 2. addition principle 3. distributive property 4. like terms

Example	Student Practice
3. Solve for x and check your solution. $9x+3=7x-2$	**4.** Solve for x and check your solution. $4x+6=x+4$

Example

3. Solve for x and check your solution.
$9x+3=7x-2$

Add $7x$ to both sides of the equation to get all variable terms on one side and combine like terms.

$$9x+3=7x-2$$
$$9x+(-7x)+3=7x+(-7x)-2$$
$$2x+3=-2$$

Add -3 to both sides and simplify.

$$2x+3=-2$$
$$2x+3+(-3)=-2+(-3)$$
$$2x=-5$$

Divide both sides by 2 and simplify.

$$2x=-5$$
$$\frac{2x}{2}=-\frac{5}{2}$$
$$x=-\frac{5}{2}$$

Check.

$$9x+3=7x-2$$
$$9\left(-\frac{5}{2}\right)+3\overset{?}{=}7\left(-\frac{5}{2}\right)-2$$
$$-\frac{45}{2}+3\overset{?}{=}-\frac{35}{2}-2$$
$$-\frac{45}{2}+\frac{6}{2}\overset{?}{=}-\frac{35}{2}-\frac{4}{2}$$
$$-\frac{39}{2}=-\frac{39}{2}$$

The solution checks.

Student Practice

4. Solve for x and check your solution.
$4x+6=x+4$

Example	Student Practice
5. Solve for x and check your solution. $5x + 26 - 6 = 9x + 12x = 25$	**6.** Solve for x and check your solution. $3x + 22 - 5 = 8x - x$

Combine like terms. Then add $(-5x)$ to both sides.

$$5x + 26 - 6 = 9x + 12x$$
$$5x + 20 = 21x$$
$$5x + (-5x) + 20 = 21x + (-5x)$$
$$20 = 16x$$

Divide both sides by 16.

$$\frac{20}{16} = \frac{16x}{16}$$
$$\frac{5}{4} = x$$

The check is left for the student.

Example	Student Practice
7. Solve for x and check your solution. $4(x+1) - 3(x-3) = 25$	**8.** Solve for x and check your solution. $-6(x-3) = 4(2x+5)$

Multiply by 4 and -3 to remove the parentheses. Then combine like terms. Be careful of the signs.

$$4(x+1) - 3(x-3) = 25$$
$$4x + 4 - 3x + 9 = 25$$
$$x + 13 = 25$$

Subtract 13 from both sides to isolate the variable.

$$x + 13 = 25$$
$$x + 13 - 13 = 25 - 13$$
$$x = 12$$

The check is left to the student.

Example	Student Practice
9. Solve for x and check your solution. $0.3(1.2x-3.6)=4.2x-16.44$	**10.** Solve for z and check your solution. $0.6(z-2)=0.4(z+7)$

Remove the parentheses.

$$0.3(1.2x-3.6)=4.2x-16.44$$
$$0.36x-1.08=4.2x-16.44$$

Solve.

$$0.36x-1.08=4.2x-16.44$$
$$0.36x+(-0.36x)-1.08=4.2x+(-0.36x)-16.44$$
$$-1.08+16.44=3.84x-16.44+16.44$$
$$15.32=3.84x$$
$$\frac{15.36}{3.84}=\frac{3.84x}{3.84}$$
$$4=x$$

The check is left to the student.

Extra Practice

1. Solve for x and check your solution.
$3x+15=30$

2. Solve for x and check your solution.
$2x-3=x+9$

3. Solve for x and check your solution.
$18-3=16x-4x+12$

4. Solve for z and check your solution.
$3(2z-3)-3=3z-2(2z-1)$

Concept Check

Explain how you would solve the equation $3(x-2)+2=2(x-4)$.

Chapter 2 Equations and Inequalities
2.4 Solving Equations with Fractions

Vocabulary
LCD • multiplication principle • infinite number of solutions • no solution • decimal

1. A(n) _____ is a fraction written a special way.

2. If there is no value of x for which an equation is true, the equation has _____.

3. To eliminate the fractions in an equation, multiply each term by the _____.

Example	**Student Practice**
1. Solve for x. $\dfrac{1}{4}x - \dfrac{2}{3} = \dfrac{5}{12}x$	**2.** Solve for x. $\dfrac{1}{18}x + \dfrac{2}{3} = \dfrac{5}{6}x$

To eliminate the fractions, multiply both sides by the LCD, 12, and apply the distributive property to simplify.

$$\frac{1}{4}x - \frac{2}{3} = \frac{5}{12}x$$

$$12\left(\frac{1}{4}x - \frac{2}{3}\right) = 12\left(\frac{5}{12}x\right)$$

$$\left(\frac{12}{1}\right)\left(\frac{1}{4}\right)x - \left(\frac{12}{1}\right)\left(\frac{2}{3}\right) = \left(\frac{12}{1}\right)\left(\frac{5}{12}\right)x$$

$$3x - 8 = 5x$$

Add $-3x$ to both sides, then divide by 2.

$$3x - 8 = 5x$$
$$3x + (-3x) - 8 = 5x + (-3x)$$
$$-8 = 2x$$
$$-\frac{8}{2} = \frac{2x}{2}$$
$$-4 = x$$

The check is left to the student.

Vocabulary Answers: 1. decimal 2. no solution 3. LCD

Example	Student Practice

Example

3. Solve for x and check your solution.

$$\frac{x}{3} + 3 = \frac{x}{5} - \frac{1}{3}$$

Multiply each term by the LCD, 15.

$$15\left(\frac{x}{3}\right) + 15(3) = 15\left(\frac{x}{5}\right) - 15\left(\frac{1}{3}\right)$$
$$5x + 45 = 3x - 5$$

Subtract $3x$ and 45 from both sides.

$$5x + 45 = 3x - 5$$
$$5x - 3x + 45 - 45 = 3x - 3x - 5 - 45$$
$$2x = -50$$

Divide by 2.

$$\frac{2x}{2} = \frac{-50}{2}$$
$$x = -25$$

Check.

$$\frac{(-25)}{3} + 3 \overset{?}{=} \frac{(-25)}{5} - \frac{1}{3}$$
$$-\frac{25}{3} + \frac{9}{3} \overset{?}{=} -\frac{5}{1} - \frac{1}{3}$$
$$-\frac{16}{3} \overset{?}{=} -\frac{15}{3} - \frac{1}{3}$$
$$-\frac{16}{3} = -\frac{16}{3}$$

The solution checks.

Student Practice

4. Solve for x and check your solution.

$$\frac{x}{2} + 2 = \frac{x}{3} - \frac{1}{2}$$

Example	Student Practice
5. Solve for x and check your solution.	**6.** Solve for x and check your solution.

5. Solve for x and check your solution.

$$\frac{1}{3}(x-2) = \frac{1}{5}(x+4) + 2$$

Remove the parentheses.

$$\frac{x}{3} - \frac{2}{3} = \frac{x}{5} + \frac{4}{5} + 2$$

Multiply all terms by the LCD, 15.
Then, solve the resulting equation.

$$15\left(\frac{x}{3}\right) - 15\left(\frac{2}{3}\right) = 15\left(\frac{x}{5}\right) + 15\left(\frac{4}{5}\right) + 15(2)$$

$$5x - 10 = 3x + 12 + 30$$

$$5x - 10 = 3x + 42$$

$$2x = 52$$

$$x = 26$$

The check is left to the student.

6. Solve for x and check your solution.

$$\frac{1}{4}(x-4) = \frac{1}{6}(x+2) + 8$$

---

7. Solve for x.

$$0.2(1 - 8x) + 1.1 = -5(0.4x - 0.3)$$

Remove parentheses. Then, multiply each term by 10 to get integer coefficients and solve.

$$0.2 - 1.6x + 1.1 = -2.0x + 1.5$$

$$2 - 16x + 11 = -20x + 15$$

$$4x + 13 = 15$$

$$4x = 2$$

$$x = \frac{1}{2} \text{ or } 0.5$$

The decimal form of the solution should only be given if it is a terminating decimal. The check is left to the student.

8. Solve for x.

$$0.3(1 - 6x) + 1.3 = 3(-0.8x + 0.6)$$

Extra Practice

1. Solve for p. Check your solution.

$$\frac{1}{3}p = p - 1$$

2. Solve for x. Check your solution.

$$\frac{x+2}{3} = \frac{x}{18} + \frac{1}{9}$$

3. Solve for x. Check your solution.

$$\frac{3}{4}(2x+6) - 3 = 3(x+3)$$

4. Solve for x. Check your solution.

$$0.4(x-6) = 0.7(2x+3) + 0.5$$

Concept Check

Explain how you would solve $\dfrac{x+5}{6} = \dfrac{x}{2} + \dfrac{3}{4}$.

Chapter 2 Equations and Inequalities
2.5 Formulas

Vocabulary

formulas   •   distance   •   rate   •   time   •   straight line

1. The formula $d = rt$ is used to find the _____ when we know the rate and time.

2. _____ are equations with one or more variables that describe real-life situations.

3. A(n) _____ can be described by an equation of the form $Ax + By = C$, where A, B, and C are real numbers and A and B are not both zero.

Example	Student Practice
1. American Airlines recently scheduled a nonstop flight in a new Boeing 777 from Chicago to London. The approximate air distance traveled on the flight was 3975 miles. The average speed of the aircraft on the trip was 530 miles per hour. How many hours did it take the Boeing 777 to fly this trip? (Source:www.aa.com)	2. An airline is planning a route from Charlotte, NC, to Los Angeles, CA. The distance between the two cities is approximately 2120 miles and the jet can do the trip in 5 hours. Find the average rate of speed for the plane.

Use the distance formula, $d = rt$ and substitute the known values.

$$d = rt$$
$$3975 = 530t$$

Divide both sides by 530 to solve for t.

$$3975 = 530t$$
$$\frac{3975}{530} = \frac{530t}{530}$$
$$7.5 = t$$

It took the jet about 7.5 hours to fly this trip from Chicago to London.

Vocabulary Answers: 1. distance 2. formulas 3. straight line

Example	Student Practice
3. Solve for t. $d = rt$	**4.** Solve for r. $C = 2\pi r$

3. Solve for t. $d = rt$

Divide both sides of the equation by the coefficient of t, which is r, to isolate t.

$$d = rt$$

$$\frac{d}{r} = \frac{rt}{r}$$

$$\frac{d}{r} = t$$

4. Solve for r. $C = 2\pi r$

---

5. Solve for y. $3x - 2y = 6$

To isolate the term containing y, subtract $3x$ from both sides.

$$3x - 2y = 6$$

$$3x - 3x - 2y = 6 - 3x$$

$$-2y = 6 - 3x$$

Divide both sides by -2.

$$-2y = 6 - 3x$$

$$\frac{-2y}{-2} = \frac{6 - 3x}{-2}$$

$$y = \frac{6 - 3x}{-2}$$

Rewrite the right hand side as two fractions, and simplify.

$$y = \frac{6 - 3x}{-2}$$

$$y = \frac{6}{-2} + \frac{-3x}{-2}$$

$$y = \frac{3x}{2} - 3$$

This is known as the slope-intercept form of the equation of a line.

6. Solve for y. $-x + 4y = -12$

Example	Student Practice

7. A trapezoid is a four-sided figure with two parallel sides. In some houses, two of the sides of the roof are in the shape of the trapezoid. If the parallel sides are a and b, and the altitude is h, the area is given by $A = \dfrac{h}{2}(a+b)$. Solve the equation for a.

Remove the parentheses by multiplying.

$$A = \frac{h}{2}(a+b)$$

$$A = \frac{ha}{2} + \frac{hb}{2}$$

Multiply all terms by the LCD, 2, and simplify.

$$A = \frac{ha}{2} + \frac{hb}{2}$$

$$2(A) = 2\left(\frac{ha}{2}\right) + 2\left(\frac{hb}{2}\right)$$

$$2A = ha + hb$$

Isolate the a term by subtracting hb from both sides.

$$2A = ha + hb$$

$$2A - hb = ha + hb - hb$$

$$2A - hb = ha$$

Divide by h, the coefficient of a.

$$2A - hb = ha$$

$$\frac{2A - hb}{h} = \frac{ha}{h}$$

$$\frac{2A - hb}{h} = a$$

8. Solve the equation $M = \dfrac{2}{5}(4y + 2x)$ for x.

Chapter 3 Solving Applied Problems
3.3 Solving Word Problems: Comparisons

Vocabulary
algebraic expression   •   comparisons   •   compare   •   check

1. Many real-life problems involve _____.

2. When solving word problems, you need to identify each quantity and write a(n) _____ that describes the situation in the word problem.

3. Word problems often _____ quantities such as length, height, or income.

Example	Student Practice
1. The Center City Animal Hospital treated a total of 18,360 dogs and cats last year. The hospital treated 1376 more dogs than cats. How many dogs were treated last year? How many cats were treated last year?	**2.** A hospital treated a total of 25,548 children and adults last year. The hospital treated 5942 more children than adults. How many children were treated last year? How many adults were treated last year?

The given information is that the combined number of dogs and cats is 18,360. The amount of dogs is being compared to the amount of cats. There were 1376 more dogs than cats. Since we are comparing the number of dogs to the number of cats, we start with the number of cats. Let $c =$ the number of cats treated at the hospital. Then $c + 1376 =$ the number of dogs treated at the hospital. So $c + (c + 1376) = 18,360$.

Solve and state the answer.
$$2c + 1376 = 18,360$$
$$2c = 16,984$$
$$c = 8492$$

8492 cats were treated and 9868 dogs were treated. The check is left to the student.

Vocabulary Answers: 1. comparisons 2. algebraic expression 3. compare

Example	Student Practice
3. An airport filed a report showing the number of plane departures from the airport during each month last year. The number of departures in March was 50 more than the number of departures in January. In July, the number of departures was 150 less than triple the number of departures in January. In those three months, the airport had 2250 departures. How many departures were recorded for each month?	**4.** An airport filed a report showing the number of plane departures from the airport during each month last year. The number of departures in May was 89 more than the number of departures in June. In November, the number of departures was 638 less than five times the number of departures in June. In those three months, the airport had 2965 departures. How many departures were recorded for each month?

Understand the problem. What is the basis of comparison?

The number of departures in March is compared to the number in January. The number of departures in July is compared to the number in January. Express this algebraically. Let $j = $ the departures in January. Then $j + 50 = $ the departures in March and $3j - 150 = $ the departures in July. Write an equation.

$$j + (j + 50) + (3j - 150) = 2250$$

Solve and state the answer.

$$j + (j + 50) + (3j - 150) = 2250$$
$$5j - 100 = 2250$$
$$5j = 2350$$
$$j = 470$$

Now, find $j + 50$ and $3j - 150$.

$$470 + 50 = 520 \text{ and } 3(470) - 150 = 1260$$

There were 470 departures in January, 520 in March, and 1260 in July. The check is left to the student.

Example	Student Practice
5. A small plot of land is in the shape of a rectangle. The length is 7 meters longer than the width. The perimeter of the rectangle is 86 meters. Find the dimensions of the rectangle.	**6.** A small plot of land is in the shape of a rectangle. The length is 11 inches longer than the width. The perimeter of the rectangle is 174 inches. Find the dimensions of the rectangle.

Understand the problem. What information is given?

The perimeter of the rectangle is 86 meters.

What is being compared?

The length is being compared to the width.

Express this algebraically and draw a picture. Let $w =$ the width. Then $w + 7 =$ the length.

Write an equation. The perimeter is the total distance around the rectangle.

$$w + (w + 7) + w + (w + 7) = 86$$

Solve and state the answer.

$$w + (w + 7) + w + (w + 7) = 86$$
$$4w + 14 = 86$$
$$4w = 72$$
$$w = 18$$

Determine the length, $18 + 7 = 25$. The width of the rectangle is 18 meters and the length is 25 meters. The check is left to the student.

Extra Practice

1. A school flag has a width that is 2 feet more than half the length. It has 28 feet of fringe sewn around its edge. What are the dimensions of the flag? Check to see if your answer is reasonable.

2. Anastasia reads 820 pages over three days. On Tuesday, she read 40 pages less than twice the number of pages she read on Monday. On Wednesday, she read 20 pages less than she read on Monday. How many pages did she read on each day? Check to see if your answer is reasonable.

3. Evan, Paul, and Raoul had a contest to see who could do more pushups. Evan did $\frac{7}{8}$ as many pushups as Paul and five fewer than Raoul. All together, the boys did 225 pushups. How many pushups did Evan do? Check to see if your answer is reasonable.

4. Mary takes care of mice in a laboratory. When she counted the mice in the morning, she found that the big cage had twice as many mice as the small cage. In the afternoon, she moved three mice from the big cage to the small cage. At that point, the big cage had six more mice than small cage. How many mice did the big cage start out with? Check to see if your answer is reasonable.

Concept Check

Explain how you would set up an equation to solve the following problem.
When Don and Laurie got married, Laurie earned $2600 more per year than Don. Together they earned $71,200 per year. How much did each of them earn for the year?

Name: _____   Date: _____
Instructor: _____   Section: _____

Chapter 3 Solving Applied Problems
3.4 Solving Word Problems: The Value of Money and Percents

Vocabulary
percent   •   simple interest   •   compound interest   •   interest

1. _____ is a charge for borrowing money or an income from investing money.

2. _____ is computed by multiplying the amount of money borrowed or invested times the rate of interest times the period of time over which it is borrowed or invested.

3. Applied situations often require finding a _____ of an unknown number.

Example	Student Practice
1. A business executive rented a car. The Supreme Car Rental Agency charged $39 per day and $0.28 per mile. The executive rented the car for two days and the total rental cost was computed to be $176. How many miles did the executive drive the rented car?	**2.** A business woman rented a room at a motel. The motel charged $52 per day and $2.50 per hour of internet use. The business woman stayed in the room for four days and the total charge was computed to be $248. How many hours of internet use did the business woman accrue?

Understand the problem. It is known that it costs $176 to rent the car for two days. It is necessary to find the number of miles the car was driven. Let $m =$ the number of miles driven in the rented car. Write an equation. Use the relationship for calculating the total cost.

per-day cost + mileage cost = total cost

$$(39)(2) \quad + \quad (0.28)m \quad = \quad 176$$

Solve and state the answer.
$$78 + 0.28m = 176$$
$$0.28m = 98$$
$$m = 350$$

The executive drove 350 miles. The check is left to the student.

Vocabulary Answers: 1. interest 2. simple interest 3. percent

Example	Student Practice

3. A sofa was marked with the following sign: "The price of this sofa has been reduced by 23%. You save $138 if you buy now." What was the original price of the sofa?

Understand the problem. Let $s =$ the original price of the sofa. Then $0.23s =$ the amount of the price reduction, which is $138.

Write an equation and solve.

$$0.23s = 138$$
$$\frac{0.23s}{0.23} = \frac{138}{0.23}$$
$$s = 600$$

The original price of the sofa was $600. The answer is reasonable because $(0.23)(600) = 138$.

4. A refrigerator was marked with the following sign: "The price of this refrigerator has been reduced by 33%. You save $264 if you buy now." What was the original price of the refrigerator?

5. Find the interest on $3000 borrowed at a simple interest rate of 18% for one year.

The simple interest formula is $I = prt$.

Substitute the values of the variables, principal $= 3000$, rate $= 18\% = 0.18$, and time $=$ one year.

$$I = prt$$
$$I = (3000)(0.18)(1)$$
$$I = 540$$

Thus, the interest charge for borrowing $3000 for one year at a simple interest rate of 18% is $540.

6. Find the interest on $5500 borrowed at a simple interest rate of 23% for one year.

Example	Student Practice

7. When Bob got out of math class, he had to make a long-distance call. He had exactly enough dimes and quarters to make a phone call that would cost $2.55. He had one fewer quarter than he had dimes. How many coins of each type did he have?

Let $d =$ the number of dimes. Then $d - 1 =$ the number of quarters. The total value of the coins was $2.55. Each dime is worth $0.10 and each quarter is worth $0.25. So, d dimes are worth $0.10d$ dollars and $(d - 1)$ quarters are worth $0.25(d - 1)$ dollars. Now write an equation for the total value, and solve.

$$0.10d + 0.25(d - 1) = 2.55$$
$$0.10d + 0.25d - 0.25 = 2.55$$
$$0.35d - 0.25 = 2.55$$
$$0.35d = 2.80$$
$$d = 8$$

Find the number of quarters Bob has, $d - 1 = 8 - 1 = 7$. Thus, Bob has eight dimes and seven quarters.

Check the answer. Bob has $8 - 7 = 1$ less quarter than he has dimes. Check that eight dimes and seven quarters are worth $2.55.

$$8(\$0.10) + 7(\$0.25) \overset{?}{=} \$2.55$$
$$\$0.80 + \$1.75 \overset{?}{=} \$2.55$$
$$\$2.55 = \$2.55$$

The solution checks.

8. When Rebecca got out of chemistry class, she went to the vending machines for a snack. She had exactly enough nickels and dimes to get a combination of items costing $2.75. She had four fewer dimes than she had nickels. How many coins of each type did she have?

Extra Practice

1. Angelina received a pay raise this year. The raise was 5% of last year's salary. This year, Angelina earned $17,640. What was her salary before the raise?

2. Find the simple interest on $6,500 borrowed at 15% for one year.

3. Randall is due to receive a 7% raise, which in dollars will be $3,780 per year. What is his current salary?

4. Mr. Finch keeps money in his pillowcase. Right now, he has equal numbers of five, ten, and twenty-dollar bills, with no other denominations. He has exactly $1505. How many bills does he have all together?

Concept Check

Explain how you would set up an equation to solve the following problem.

Robert has $2.55 in change consisting of nickels, dimes, and quarters. He has twice as many dimes as quarters. He has one more nickel than he has quarters. How many of each coin does he have?

Name: _____   Date: _____
Instructor: _____   Section: _____

Chapter 3 Solving Applied Problems
3.5 Solving Word Problems Using Geometric Formulas

Vocabulary
equilateral • isosceles • right • area • perimeter • volume • surface area
sphere • rectangular prism • right circular cylinder

1. The _____ of a three-dimensional figure is the measure of the amount of space inside the figure.

2. The _____ of a sphere is given by the formula $4\pi r^2$.

3. A(n) _____ triangle is a triangle with one angle that measures $90°$.

4. A(n) _____ triangle is a triangle with two equal sides. The two angles opposite the equal sides are also equal.

Example	Student Practice
1. Find the area of a triangular window whose base is 16 inches and whose altitude is 19 inches.	**2.** Find the area of a triangular table whose base is 120 centimeters and whose altitude is 150 centimeters.

Use the formula for the area of a triangle. Substitute the known values in the formula and solve for the unknown.

$A = \frac{1}{2}ab$

$= \frac{1}{2}(16 \text{ in.})(19 \text{ in.})$

$= \frac{1}{2}(16)(19)(\text{in.})(\text{in.})$

$= 152 \text{ in.}^2$

The area of the triangle is 152 square inches.

Vocabulary Answers: 1. volume 2. surface area 3. right 4. isosceles

Example	Student Practice
3. Find the perimeter of a parallelogram whose longer sides are 4 feet and whose shorter sides are 2.6 feet.	**4.** Find the perimeter of a parallelogram whose longer sides are 5.7 miles and whose shorter sides are 3 miles.

3. Draw a picture. The perimeter is the distance around the figure. To find the perimeter, add the lengths of the sides. Since the opposite sides of a parallelogram are equal, we can write the following.

$$P = 2(4 \text{ feet}) + 2(2.6 \text{ feet})$$
$$= 8 \text{ feet} + 5.2 \text{ feet}$$
$$= 13.2 \text{ feet}$$

The perimeter is 13.2 feet.

Example	Student Practice
5. The smallest angle of an isosceles triangle measures $24°$. The other two angles are larger. What are the measurements of the other two angles?	**6.** The smallest angle of an isosceles triangle measures $58°$. The other two angles are larger. What are the measurements of the other two angles?

In an isosceles triangle, the measures of two angles are equal. We know that the sum of the measures of all three angles is $180°$. Both of the larger angles must be different from $24°$, therefore these two larger angles must be equal. Let $x =$ the measure in degrees of each of the larger angles.

$$24° + x + x = 180°$$
$$24° + 2x = 180°$$
$$2x = 156°$$
$$x = 78°$$

Thus, the measures of each of the other two angles of the triangle must be $78°$.

Example	Student Practice
7. Find the volume of a sphere with radius 4 centimeters. (Use 3.14 for π.) Round your answer to the nearest cubic centimeter.	**8.** Find the volume of a sphere with radius 19 feet. (Use 3.14 for π.) Round your answer to the nearest cubic centimeter.

$$V = \frac{4}{3}\pi r^3$$

$$= \frac{4}{3}(3.14)(4 \text{ cm})^3$$

$$= \frac{4}{3}(3.14)(64) \text{ cm}^3$$

$$\approx 268 \text{ cm}^3$$

9. A can is made of aluminum. It has a flat top and a flat bottom. The height is 5 inches and the radius is 2 inches. How much aluminum is needed to make the can? How much aluminum is needed to make 10,000 cans?

Use the formula for the surface area of a right circular cylinder. Use 3.14 to approximate π.

$$SA = 2\pi rh + 2\pi r^2$$

$$= 2(3.14)(2)(5) + 2(3.14)(2)^2$$

$$= 62.8 + 24.12 = 87.92 \text{ in.}^2$$

87.92 in.^{2} of aluminum are needed to make each can. Determine how much aluminum is needed to make 10,000 cans.

$$(87.92 \text{ in.}^2)(10,000) = 879,200 \text{ in.}^2$$

It would take 879,200 square inches of aluminum to make 10,000 cans.

10. A tube is made of cardboard. It has a flat top and a flat bottom. The height is 30 inches and the radius is 3 inches. How much cardboard is needed to make the tube? How much cardboard is needed to make 8,000 tubes? (Use 3.14 for π.)

107
Copyright © 2013 Pearson Education, Inc.

Example	Student Practice

11. A quarter-circle (of radius 1.5 yards) is connected to two rectangles with dimensions as labeled on the sketch below. You need to lay carpet in your house according to this sketch. How many square yards of carpeting will be needed? (Use 3.14 for π.) Round your final answer to the nearest tenth.

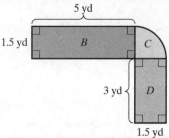

12. A quarter-circle (of radius 3.75 feet) is connected to two rectangles with dimensions as labeled on the sketch below. You need to lay turf at a miniature golf course according to this sketch. How many square feet of turf will be needed? (Use 3.14 for π.) Round your final answer to the nearest tenth.

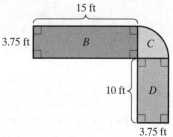

The desired area is the sum of the three areas, which we will call B, C, and D. Area B and Area D are rectangular in shape. Find the areas of these shapes.

$$A_B = (5 \text{ yd})(1.5 \text{ yd}) = 7.5 \text{ yd}^2$$
$$A_D = (3 \text{ yd})(1.5 \text{ yd}) = 4.5 \text{ yd}^2$$

Area C is one-fourth of a circle. The radius of the circle is 1.5 yd.

$$A_C = \frac{\pi r^2}{4}$$
$$= \frac{(3.14)(1.5 \text{ yd})^2}{4}$$
$$\approx 1.8 \text{ yd}^2$$

The total area is as follows.

$$A_B + A_D + A_C \approx 7.5 + 4.5 + 1.8$$
$$\approx 13.8$$

13.8 yd^2 of carpeting is needed.

Extra Practice

1. What is the area of a parallelogram with a base of 12 cm and an altitude of 5 cm?

2. What are the surface area and volume of a sphere with a radius of 22 inches? (Use 3.14 for π.) Round to the nearest whole number.

3. The perimeter of a trapezoid is 25 inches. The two sides and the shorter base are all the same length, while the longer base is twice this length. What is the length of the longer base?

4. A cylindrical storage container is 10 meters tall and has a capacity of 1538.6 cubic meters. If you walk at a rate of 2 meters per second, how much time (to the nearest whole number) will you need to walk all the way around the container?

Concept Check

Explain how you would set up an equation to solve the following problem.

Two angles of a triangle measure 135° and 11°. What is the measure of the third angle?

Chapter 3 Solving Applied Problems
3.6 Using Inequalities to Solve Word Problems

Vocabulary
greater than • less than • greater than or equal to • less than or equal to

1. The phrases "is at least" and "cannot be less than" mean "_____."

2. The symbol for "_____" is ≤.

3. The phrase "is smaller than" means "_____."

4. The symbol for "_____" is >.

Example	Student Practice
1. Translate each sentence into an inequality.	**2.** Translate each sentence into an inequality.
(a) The perimeter of the rectangle must be no more than 70 inches.	**(a)** The area of the field must be no less than 120 square yards.
"no more than" indicates "less than or equal to."	
Perimeter ≤ 70	
(b) The average number of errors must be less than 2 mistakes per page.	**(b)** Average grade must be greater than 65 to pass.
Average < 2	
	(c) Bill total must be at most $1570.
(c) His income must be at least $950 per week.	
Income ≥ 950	
(d) The new car will cost at most $12,070.	**(d)** The new phone will cost at least $400.
Car ≤ 12,070	

Vocabulary Answers: 1. greater than or equal to 2. less than or equal to 3. less than 4. greater than

Example	Student Practice

3. A manufacturing company makes 60-watt fluorescent light bulbs. For every 1000 bulbs manufactured, a random sample of five bulbs is selected and tested. To meet quality control standards, the average number of hours these five bulbs last must be at least 900 hours. The test engineer tested five bulbs. Four of them lasted 850 hours, 1050 hours, 1000 hours, and 950 hours, respectively. How many hours must the fifth bulb last in order for the batch to meet quality control standards?

Let n = the number of hours the fifth bulb must last. The average that the five bulbs last must be at least 900 hours. Thus, Average ≥ 900 hours. Write an inequality.

$$\text{Average of five bulbs} \geq 900 \text{ hours}$$

$$\frac{850+1050+1000+950+n}{5} \geq 900$$

Solve and state the answer.

$$\frac{850+1050+1000+950+n}{5} \geq 900$$

$$\frac{3850+n}{5} \geq 900$$

$$3850+n \geq 4500$$

$$n \geq 650$$

If the fifth bulb burns for 650 hours or more, the quality control standards will be met. The check is left to the student.

4. A manufacturing company makes cell phone batteries. For every 1000 batteries manufactured, a random sample of five batteries is selected and tested. To meet quality control standards, the average number of hours these batteries last must be at least 8.75 hours. The test engineer tested five batteries. Four of them lasted 8 hours, 9.5 hours, 7.75 hours, and 6 hours, respectively. How many hours must the fifth battery last in order for the batch to meet quality control standards?

Example	Student Practice

5. Juan is selling supplies for restaurants in the Southwest. He gets paid $700 per month plus 8% of the value of all the supplies he sells to restaurants. If he wants to earn more than $3100 per month, what value of products will he need to sell?

Juan earns a fixed salary and a salary that depends on how much he sells (commission salary) each month. Let $x =$ the value of the supplies he sells each month. Then $0.08x =$ the amount of commission salary he earns each month. He wants his total income to be more than $3100 per month. Write the inequality. The fixed income added to the commission income must be greater than $3100.

$$700 + 0.08x > 3100$$

Solve and state the answer.

$$700 + 0.08x > 3100$$
$$0.08x > 2400$$
$$x > 30,000$$

Juan must sell more than $30,000 worth of supplies each month. Check the answer. Any value greater than 30,000 should satisfy the original inequality.

$$700 + 0.08(30,001) \overset{?}{>} 3100$$

$$700 + 2400.08 \overset{?}{>} 3100$$

$$3100.08 > 3100$$

The solution checks.

6. Aasif is selling supplies for restaurants along the western seaboard. He gets paid $1500 per month plus 5% of the value of all the supplies he sells to restaurants. If he wants to earn more than $6500 per month, what value of products will he need to sell?

113

Extra Practice

1. Translate to an inequality.

   The height of the box is less than 4 feet.

2. Translate to an inequality.

   To make money, the theater must fill at least 120 seats.

3. Jackson wants to plant a rectangular vegetable garden beside his house. He wants the garden to be at least 42 square feet in area but only 3 feet wide. What possible lengths for the garden satisfy the given conditions?

4. Alicia earns extra money from babysitting. She likes to average $100 per week over any four week period. In the last three weeks, she earned $70, $135, and $85. How much must she earn this week to meet her goal of averaging $100 per week?

Concept Check

Explain how you would set up an inequality to solve the following problem.
Caleb wants to spend no more than $150 to rent a truck to move things for his dad's company. The rental company charges $37.50 per day to rent the truck, and 19.5 cents per mile that the truck is driven. If he rented the truck for two days, how many miles could he drive the truck?

MATH COACH

Mastering the skills you need to do well on the test.

Watch the **MATH COACH** videos in MyMathLab® or on You[Tube]™ while you work the problems below. These helpful hints will help you avoid making common errors on test problems.

Solving Applied Problems Involving Comparisons—

Problem 5   A rectangle has a length 7 meters longer than double the width. The perimeter is 134 meters. Find the dimensions of the rectangle.

> **Helpful Hint:** When one quantity is being compared to another quantity, always pick as your starting variable the quantity that is being compared to.

Did you first let w = the width since the length is compared to the width? Did you then write $2w + 7$ for the length? Yes _____ No _____

If you answered No, remember that the rectangle has a length 7 meters longer than double the width. We need to start with the width. Since the length is 7 meters longer than double the width, we write $2w + 7$ as the length.

Did you then write the equation for the perimeter? $P = 2w + 2l$. Thus we would write $2(w) + 2(2w + 7) = 134$. This becomes $2w + 4w + 14 = 134$.

Finally we obtain $6w + 14 = 134$. Were you able to write that equation? Yes _____ No _____

If you answered No, remember that we must have double the width plus double the length equals the perimeter of 134.

If you answered Problem 5 incorrectly, rework it now using these suggestions.

Solving Applied Problems Involving Money and Percents—Problem 9
Franco invested $4000 in money market funds. Part was invested at 14% interest, the rest at 11% interest. At the end of each year the fund company pays interest. After one year he earned $482 in simple interest. How much was invested at each interest rate?

> **Helpful Hint:** Let x = the amount of one investment and then the $(\text{total} - x)$ = the amount of the other investment. Then multiply each investment by the respective interest rate.

Let us suppose that you picked x as the amount of money that was invested at 14% interest. Did you realize that you should use $(4000 - x)$ to represent the amount of money invested at 11%? Yes _____ No _____

If you answered No, remember that we want the $(\text{total} - x)$ to represent the second amount of money. The two amounts together have to add up to $4000. That is why we use x for one amount and $(4000 - x)$ to represent the other amount.

Were you able to write the equation $0.14x + 0.11(4000 - x) = 482$? Can you simplify that equation to obtain $0.14x + 440 - 0.11x = 482$?

Do you realize that this becomes the equation $0.03x + 440 = 482$? Yes _____ No _____

If you answered No, remember that we must multiply x by 14%, which is $0.14x$. We must multiply $(4000 - x)$ by 11%, which is $0.11(4000 - x)$. We must be very careful when we remove the parentheses to get the simplified equation.

If you answered Problem 9 incorrectly, rework it now using these suggestions.

Solving Applied Problems Using Geometric Formulas—Problem 15 How much would it cost to carpet the area Shown in the figure if carpeting costs $12 per square yard?

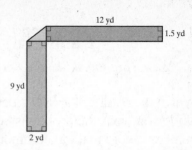

Helpful Hint: Consider that the figure consists of three separate parts. Two are rectangles and one is a triangle. Find the area of each separate part. Then add these smaller areas in order to obtain the total area.

Did you find the area of the top rectangle to be $12 \times 1.5 = 18$ square yards? Did you find the area of the bottom rectangle to be $9 \times 2 = 18$ square yards?

Yes _____ No _____

If you answered No, please go back and look at each rectangle separately. For each one you must multiply the length times the width. Be sure to label your answer as square yards.

Did you realize that the triangle has an altitude of 1.5 yards and a base of 2 yards?

Did you multiply $\left(\frac{1}{2}\right)(1.5)(2)$ and obtain the area of the triangle of 1.5 square yards?

Did you add up the three areas to obtain 37.5 square yards?

Yes _____ No _____

If you answered No, please go back and examine the dimensions of the triangle carefully. Although they are not labeled on the figure, they can be quickly determined by looking at the other side of each rectangle.

Then use the area formula $A = \left(\frac{1}{2}\right)(bh)$. Then find the total of the three areas.

If you answered Problem 15 incorrectly, rework it now using these suggestions.

Solving Applied Problems Using Inequalities—Problem 17 Carol earns $15,000 plus 5% commission on sales over $10,000. How much must she make in sales to earn more than $20,000?

Helpful Hint: The commission income is base on how much above $10,000 she obtains from sales. So if we let $x =$ the amount of sales she needs, then she is only earning commission on $(x - 10,000)$ dollars.

Did you realize that the expression for how much she earns from commission sales is $0.05(x - 10,000)$, where x is the amount of sales in dollars?

Yes _____ No _____

If you answered No, remember that we must find 5% of the amount of money that she obtains in sales that is above $10,000. Therefore we have the expression $0.05(x - 10,000)$.

Did you obtain the final inequality of her earnings to be $15,000 + 0.05(x - 10,000) > 20,000$ since she wants to earn more than $20,000$?

Can you simplify this inequality to obtain $15,000 + 0.05x - 500 > 20,000$?

Can you further simplify this to obtain $14,500 + 0.05x > 20,000$?

Yes _____ No _____

If you answered No, remember that we must add her yearly salary plus her commission salary and that this must be greater than $20,000.

If you answered Problem 17 incorrectly, rework it now using these suggestions.

Name: _____   Date: _____
Instructor: _____   Section: _____

Chapter 4 Exponents and Polynomials
4.1 The Rules of Exponents

Vocabulary
exponent • base • exponential expression • the product rule
numerical coefficient • the quotient rule

1. In the exponential expression x^a, x is called the _____.

2. In the exponential expression x^a, a is called the _____.

3. Simplifying the exponential expression $\dfrac{x^a}{x^b}$ requires using the _____.

4. A(n) _____ is a number that is multiplied by a variable, such as the 4 in $4x^2$.

Example	Student Practice
1. Multiply.	**2.** Multiply.
(a) $x^3 \cdot x^6$	**(a)** $z^4 \cdot z$
The expressions have the same base so we can add the exponents.	
$x^3 \cdot x^6 = x^{3+6} = x^9$	
(b) $x \cdot x^5$	**(b)** $2^2 \cdot 2^6$
Every variable that does not have a written exponent is understood to have an exponent of 1. Thus, $x = x^1$.	
$x \cdot x^5 = x^{1+5} = x^6$	

Vocabulary Answers: 1. base 2. exponent 3. quotient rule 4. numerical coefficient

Example	Student Practice
3. Multiply. $(5ab)\left(-\dfrac{1}{3}a\right)(9b^2)$	**4.** Multiply. $(-4x^3)(xy^2)(2x^2y)$

Multiply the numerical coefficients and group like bases.

$$(5ab)\left(-\frac{1}{3}a\right)(9b^2)$$

$$= (5)\left(-\frac{1}{3}\right)(9)(a \cdot a)(b \cdot b^2)$$

Use the rule for multiplying expression with exponents. Add the exponents.

$$(5)\left(-\frac{1}{3}\right)(9)(a \cdot a)(b \cdot b^2) = -15a^2b^3$$

5. Divide. $\dfrac{2^{16}}{2^{11}}$	**6.** Divide. $\dfrac{x^9}{x^5}$

The expressions have the same base so we can subtract the exponents.

$$\frac{2^{16}}{2^{11}} = 2^{16-11} = 2^5$$

7. Divide. $\dfrac{12^{17}}{12^{20}}$	**8.** Divide. $\dfrac{n^6}{n^{10}}$

The expressions have the same base so we can subtract the exponents. Notice that the larger exponent is in the denominator.

$$\frac{12^{17}}{12^{20}} = \frac{1}{12^{20-17}} = \frac{1}{12^3}$$

Example	Student Practice
9. Simplify. $\dfrac{4x^0y^2}{8^0y^5z^3}$	**10.** Simplify. $\dfrac{\left(2y^4\right)\left(3x^2y\right)}{12x^0y^{10}}$

Any number (except 0) to the 0 power equals 1.

$$\frac{4x^0y^2}{8^0y^5z^3} = \frac{4(1)y^2}{(1)y^5z^3}$$

$$= \frac{4y^2}{y^5z^3}$$

$$= \frac{4}{y^3z^3}$$

---

11. Simplify.

    (a) $\left(x^3\right)^5$

        We are raising a power to a power, so multiply exponents.

        $\left(x^3\right)^5 = x^{3\cdot5} = x^{15}$

    (b) $\left(-1\right)^8$

        Since n is even, a positive number results.

        $\left(-1^8\right) = +1$

12. Simplify.

    (a) $\left(y^8\right)^4$

    (b) $\left(-1\right)^{13}$

---

13. Simplify. $\left(\dfrac{x}{y}\right)^5$

The fraction within the parentheses is raised to a power. Raise both the numerator and the denominator to that power.

$$\left(\frac{x}{y}\right)^5 = \frac{x^5}{y^5}$$

14. Simplify. $\left(\dfrac{y}{z^2}\right)^3$

Example	Student Practice
15. Simplify. $\left(\dfrac{-3x^2z^0}{y^3}\right)^4$	**16.** Simplify. $\left(\dfrac{4x^5y}{-16x^0y^3}\right)^3$

Simplify inside the parentheses first. Apply the rules for raising a power to a power and simplify.

$$\left(\frac{-3x^2z^0}{y^3}\right)^4 = \left(\frac{-3x^2}{y^3}\right)^4$$

$$= \frac{(-3)^4 x^8}{y^{12}} = \frac{81x^8}{y^{12}}$$

Extra Practice

1. Multiply. Leave your answer in exponent form. $\left(12x^4y^2\right)\left(2x^2y^3\right)$

2. Simplify. Leave your answer in exponent form. Assume that all variables in any denominator are nonzero. $\dfrac{9a^3}{a^3}$

3. Simplify. Leave your answer in exponent form. Assume that all variables in any denominator are nonzero. $\left(\dfrac{b}{b^3}\right)^2$

4. Simplify. Leave your answer in exponent form. Assume that all variables in any denominator are nonzero. $\left(\dfrac{9x^2y^0}{14x^5}\right)^2$

Concept Check

Explain the steps you would need to follow to simplify the expression. $\dfrac{\left(4x^3\right)^2}{\left(2x^4\right)^3}$

Chapter 4 Exponents and Polynomials
4.2 Negative Exponents and Scientific Notation

Vocabulary
exponent • base • negative exponent • scientific notation • significant digits

1. All digits in a number, excluding the zeros that allow the decimal point to be properly located, are considered _____.

2. In the number 1.23×10^5, "10" is referred to as the _____ .

3. A useful way to express very large or very small numbers is to use _____ .

4. When using scientific notation to express a number less than zero, it is necessary to use a _____ on the base of ten.

Example	Student Practice
1. Evaluate. 3^{-2}	**2.** Evaluate. 7^{-4}
First write the expression with a positive exponent. Then evaluate.	
$3^{-2} = \dfrac{1}{3^2} = \dfrac{1}{9}$	
3. Simplify. Write the expression with no negative exponents. $\left(3x^{-4}y^2\right)^{-3}$	**4.** Simplify. Write the expression with no negative exponents. $\left(2xy^{-2}z^3\right)^{-4}$
Use the power to power rule.	
$\left(3x^{-4}y^2\right)^{-3} = 3^{-3}x^{12}y^{-6}$	
Now write the expression with positive exponents only and simplify.	
$3^{-3}x^{12}y^{-6} = \dfrac{x^{12}}{3^3 y^6} = \dfrac{x^{12}}{27 y^6}$	

Vocabulary Answers: 1. significant digits 2. base 3. scientific notation 4. negative exponent

Example	Student Practice
5. Write in scientific notation. $157,000,000$	**6.** Write in scientific notation. 2564

5. Write in scientific notation. $157,000,000$

Move the decimal point 8 places to the left and multiply by $100,000,000$.

$$157,000,000 = 1.\underset{\text{8 places}}{\underline{57000000}} \times 1\underbrace{00000000}_{\text{8 zeros}}$$

$$= 1.57 \times 10^8$$

6. Write in scientific notation. 2564

---

7. Write in scientific notation.

    (a) 0.061

        Move the decimal point 2 places to the right and multiply by 10^{-2}.

        $0.061 = 6.1 \times 10^{-2}$

    (b) 0.000052

        $0.000052 = 5.2 \times 10^{-5}$

8. Write in scientific notation.

    (a) 0.0013

    (b) 0.000001

---

9. Write in decimal notation.

    (a) 1.568×10^2

        The exponent 2 tells us to move the decimal point 2 places to the right.

        $1.568 \times 10^2 = 156.8$

    (b) 7.432×10^{-3}

        The exponent -3 tells us to move the decimal point 3 places to the left.

        $7.432 \times 10^{-3} = 7.432 \times \dfrac{1}{1000}$

                    $= 0.007432$

10. Write in decimal notation.

    (a) 5.28×10^{-4}

    (b) 3.2214×10^6

Example	Student Practice

11. The approximate distance from Earth to the star Polaris is 208 parsecs (pc). A parsec is a distance of approximately 3.09×10^{13} km. How long would it take a space probe traveling at 40,000 km/hr to reach the star? Round to three significant digits.

Understand the problem. We need to change the distance from parsecs to kilometers using the given relationship.

$$208 \text{ pc} = \frac{(208 \text{ pc})(3.09 \times 10^{13} \text{ km})}{1 \text{ pc}}$$

$$= 642.72 \times 10^{13}$$

Write an equation using the distance formula, $\text{distance} = \text{rate} \times \text{time}$, or $d = r \times t$.

Substitute known values and change all given values to scientific notation.

$$6.4272 \times 10^{15} \text{ km} = \frac{4 \times 10^4 \text{ km}}{1 \text{ hr}} \times t$$

Multiply both sides by the reciprocal of $\dfrac{4 \times 10^4 \text{ km}}{1 \text{ hr}}$ and simplify.

$$6.4272 \times 10^{15} \text{ km} \times \frac{1 \text{ hr}}{4 \times 10^4 \text{ km}} = t$$

$$\frac{\left(6.4272 \times 10^{15} \text{ k\!m}\right)(1 \text{ hr})}{4 \times 10^4 \text{ k\!m}} = t$$

$$1.6068 \times 10^{11} \text{ hr} = t$$

Round to three significant figures.
$$1.6068 \times 10^{11} \text{ hr} \approx 1.61 \times 10^{11} \text{ hr}$$

The check is left to the student.

12. Use the information in example **15** to answer the following. How long would it take the space probe to reach a star that is 500 parsecs from Earth? Round to three significant digits.

123

Extra Practice

1. Simplify. Express your answer with positive exponents. Assume that all variables are nonzero. 7^{-3}

2. Simplify. Express your answer with positive exponents. Assume that all variables are nonzero. $\left(\dfrac{3a^3b^{-2}}{c^3}\right)^{-4}$

3. Write in decimal notation. 6.34×10^3

4. Evaluate by using scientific notation and the laws of exponents. Leave your answer in scientific notation. $\dfrac{0.0046}{0.023}$

Concept Check

Explain how you would simplify a problem like the following so that your answer has only positive exponents. $\left(4x^{-3}y^4\right)^{-3}$

Name: _____   Date: _____
Instructor: _____   Section: _____

Chapter 4 Exponents and Polynomials
4.3 Fundamental Polynomial Operations

Vocabulary
polynomial • multivariable polynomial • degree of a term • degree of a polynomial
monomial • binomial • trinomial • decreasing order • evaluate

1. A(n) _____ is a polynomial with two terms.

2. In a term, the sum of the exponents of all of the variables is called the _____.

3. When a polynomial is written in _____, the value of each exponent decreases as we move from left to right.

4. The highest degree of all of the terms in a polynomial is called the _____.

Example	Student Practice
1. State the degree of the polynomial, and whether it is a monomial, a binomial, or a trinomial.	**2.** State the degree of the polynomial, and whether it is a monomial, a binomial, or a trinomial.
(a) $5xy + 3x^3$	**(a)** $7y^3z^4$
This polynomial is of degree 3. It has two terms, so it is a binomial.	
(b) $-7a^5b^2$	**(b)** $x^2 + 5x - 4$
The sum of the exponents is $5 + 2 = 7$. Therefore, this polynomial is of degree 7. It has one term, so it is a monomial.	
(c) $8x^4 - 9x - 15$	**(c)** $2xy^6 + 7x^2$
This polynomial is of degree 4. It has three terms, so it is a trinomial.	

Vocabulary Answers: 1. binomial 2. degree of a term 3. decreasing order 4. degree of a polynomial

125
Copyright © 2013 Pearson Education, Inc.

Example	Student Practice
3. Add. $\left(5x^2 - 6x - 12\right) + \left(-3x^2 - 9x + 5\right)$	**4.** Add. $\left(-x^2 + x - 5\right) + \left(5x^2 - 9x + 10\right)$

Group like terms.

$$\left(5x^2 - 6x - 12\right) + \left(-3x^2 - 9x + 5\right)$$
$$= \left[5x^2 + \left(-3x^2\right)\right] + \left[-6x + (-9x)\right] + \left[-12 + 5\right]$$

Add like terms.

$$\left[(5-3)x^2\right] + \left[(-6-9)x\right] + \left[-12+5\right]$$
$$= 2x^2 + (-15x) + (-7)$$
$$= 2x^2 - 15x - 7$$

| **5.** Add. $\left(\dfrac{1}{2}x^2 - 6x + \dfrac{1}{3}\right) + \left(\dfrac{1}{5}x^2 - 2x - \dfrac{1}{2}\right)$ | **6.** Add. $\left(\dfrac{3}{4}x^2 - 2x + \dfrac{1}{8}\right) + \left(x^2 + \dfrac{2}{5}x + \dfrac{1}{2}\right)$ |

The numerical coefficients of polynomials may be any real number. Thus, polynomials may have numerical coefficients that are decimals or fractions.

To add, first group like terms.

$$\left(\frac{1}{2}x^2 - 6x + \frac{1}{3}\right) + \left(\frac{1}{5}x^2 - 2x - \frac{1}{2}\right)$$
$$= \left[\frac{1}{2}x^2 + \frac{1}{5}x^2\right] + \left[-6x + (-2x)\right] + \left[\frac{1}{3} + \left(-\frac{1}{2}\right)\right]$$

Add like terms.

$$\left[\left(\frac{1}{2} + \frac{1}{5}\right)x^2\right] + \left[(-6-2)x\right] + \left[\frac{1}{3} + \left(-\frac{1}{2}\right)\right]$$
$$= \left[\left(\frac{5}{10} + \frac{2}{10}\right)x^2\right] + (-8x) + \left[\frac{2}{6} - \frac{3}{6}\right]$$
$$= \frac{7}{10}x^2 - 8x - \frac{1}{6}$$

Example	Student Practice
7. Subtract. $\left(7x^2 - 6x + 3\right) - \left(5x^2 - 8x - 12\right)$	**8.** Subtract. $\left(-5x^2 - 3x + 1\right) - \left(-2x^2 + 7x - 4\right)$

We change the sign of each term in the second polynomial and then add.

$$\left(7x^2 - 6x + 3\right) - \left(5x^2 - 8x - 12\right)$$

$$= \left(7x^2 - 6x + 3\right) + \left(-5x^2 + 8x + 12\right)$$

$$= (7 - 5)x^2 + (-6 + 8)x + (3 + 12)$$

$$= 2x^2 + 2x + 15$$

---

9. Automobiles sold in the United States have become more fuel efficient over the years due to regulations from Congress. The number of miles per gallon obtained by the average automobile in the United States can be described by the polynomial $0.3x + 12.9$, where x is the number of years since 1970. (*Source:* U.S. Federal Highway Administration.) Use this polynomial to estimate the number of miles per gallon obtained by the average automobile in 1972.

The year 1972 is two years later than 1970, so $x = 2$.

Thus, the number of miles per gallon obtained by the average automobile in 1972 can be estimated by evaluating $0.3x + 12.9$ when $x = 2$.

$$0.3(2) + 12.9 = 0.6 + 12.9$$

$$= 13.5$$

We estimate that the average car in 1972 obtained 13.5 miles per gallon.

10. Use the information in example **9** to answer the following. Estimate the number of miles per gallon that will be obtained by the average automobile in the United States in 2015.

Extra Practice

1. State the degree of the polynomial and whether it is a monomial, a binomial, or a trinomial. $23x^6 - 14x^3 + 5$

2. Add. $(6.4x - 3) + (4.4x - 11)$

3. Subtract.

   $(2r^4 - 3r^2 + 14) - (-3r^4 - 2r^2 + 6)$

4. Buses sold in the U.S. have become more fuel efficient over the years. The number of miles per gallon obtained by a certain type of bus can be described by the polynomial $0.37x + 5.31$, where x is the number of years since 1970. Use this polynomial to estimate the number of miles per gallon obtained by this type of bus in 1980.

Concept Check

Explain how you would determine the degree of the following polynomial and how you would decide if it is a monomial, a binomial, or a trinomial. $2xy^2 - 5x^3y^4$

Name: _____   Date: _____
Instructor: _____   Section: _____

Chapter 4 Exponents and Polynomials
4.4 Multiplying Polynomials

Vocabulary
polynomial • monomial • binomial • distributive property • FOIL

1. A _____ is a polynomial with only one term.

2. The process of using the distributive property to multiply two binomials is often referred to as _____.

3. The _____ states that for all real numbers a, b, and c, $a(b+c) = ab + ac$.

Example	Student Practice
1. Multiply. $3x^2(5x-2)$	**2.** Multiply. $5y^2(-y+4)$

Use the distributive property and multiply each term by $3x^2$.

$$3x^2(5x-2) = 3x^2(5x) + 3x^2(-2)$$

Simplify.

$$3x^2(5x) + 3x^2(-2)$$
$$= (3\cdot5)(x^2\cdot x) + (3)(-2)x^2$$
$$= 15x^3 - 6x^2$$

3. Multiply. $(x^2-2x+6)(-2xy)$ **4.** Multiply. $(x^2-7x-10)(2x^2y)$

Use the distributive property and multiply each term by $-2xy$.

$$(x^2-2x+6)(-2xy)$$
$$= -2x^3y + 4x^2y - 12xy$$

Vocabulary Answers: 1. monomial 2. FOIL 3. distributive property

Example	Student Practice
5. Multiply. $(2x-1)(3x+2)$	**6.** Multiply. $(3x-1)(5x+3)$

5. Multiply. $(2x-1)(3x+2)$

Multiply the First terms, $2x$ and $3x$.
$(2x)(3x)=6x^2$

Multiply the Outer terms, $2x$ and 2.
$(2x)(2)=4x$

Multiply the Inner terms, -1 and $3x$.
$(-1)(3x)=-3x$

Multiply the Last terms, -1 and 2.
$(-1)(2)=-2$

Add the results and combine like terms.
  First + Outer + Inner + Last

$= 6x^2 \ + \ 4x \ - \ 3x \ - \ 2$

$= 6x^2 + x - 2$

6. Multiply. $(3x-1)(5x+3)$

---

7. Multiply. $(3x+2y)(5x-3z)$

Multiply the First terms, $3x$ and $5x$.
$(3x)(5x)=15x^2$

Multiply the Outer terms, $3x$ and $-3z$.
$(3x)(-3z)=-9xz$

Multiply the Inner terms, $2y$ and $5x$.
$(2y)(5x)=10xy$

Multiply the Last terms, $2y$ and $-3z$.
$(2y)(-3z)=-6yz$

Add the results and combine like terms.
$(3x+2y)(5x-3z)$

$= 15x^2 - 9xz + 10xy - 6yz$

8. Multiply. $(7x-3y)(x+2z)$

Example	Student Practice
9. Multiply. $(7x-2y)^2$	**10.** Multiply. $(4x+3y)^2$

When we square a binomial, it is the same as multiplying the binomial by itself.

$$(7x-2y)(7x-2y)$$

Multiply the First terms, $7x$ and $7x$.
$$(7x)(7x)=49x^2$$

Multiply the Outer terms, $7x$ and $-2y$.
$$(7x)(-2y)=-14xy$$

Multiply the Inner terms, $-2y$ and $7x$.
$$(-2y)(7x)=-14xy$$

Multiply the Last terms, $-2y$ and $-2y$.
$$(-2y)(-2y)=4y^2$$

Add the results and combine like terms.

$$(7x-2y)^2 = 49x^2 -14xy -14xy +4y^2$$
$$= 49x^2 - 28xy + 4y^2$$

11. Multiply. $\left(3x^2 +4y^3\right)\left(2x^2 +5y^3\right)$	**12.** Multiply. $\left(10x^2 +5y^4\right)\left(2x^2 -3y^4\right)$

Use the FOIL method and the rules for multiplying expressions with exponents.

$$\left(3x^2 +4y^3\right)\left(2x^2 +5y^3\right)$$
$$= 6x^4 +15x^2 y^3 +8x^2 y^3 +20y^6$$
$$= 6x^4 +23x^2 y^3 +20y^6$$

Example	Student Practice
13. The width of a living room is $(x+4)$ feet. The length of the room is $(3x+5)$ feet. What is the area of the room in square feet?	**14.** The width of a brick patio is $(2x+1)$ feet. The length of the patio is $(4x-2)$ feet. What is the area of the patio in square feet?

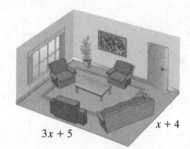

$3x + 5$ $x + 4$

Use the area formula and solve.

$$A = (\text{length})(\text{width})$$
$$A = (3x+5)(x+4)$$
$$A = 3x^2 + 12x + 5x + 20$$
$$A = 3x^2 + 17x + 20$$

There are $\left(3x^2 + 17x + 20\right)$ square feet in the room.

Extra Practice

1. Multiply. $5x\left(-2x^3 + 3x\right)$

2. Multiply. $(x-3)(x-11)$

3. Multiply. $(3x-7y)(5x+8y)$

4. Multiply. $\left(2x^2 - 3y^3\right)\left(4x^2 + 5y^3\right)$

Concept Check

Explain how you would multiply $(7x-3)^2$.

Chapter 4 Exponents and Polynomials
4.5 Multiplication: Special Cases

Vocabulary
polynomial   •   term   •   binomial   •   FOIL   •   square of a sum
square of a difference   •   vertical multiplication   •   horizontal multiplication

1. One method of multiplying polynomials that uses a method similar to that used in arithmetic for multiplying whole numbers is called _____.

2. A binomial of the form $(a-b)^2$ is referred to as the _____.

3. One way to multiply polynomials with more than two terms is to use _____ horizontal multiplication.

Example	Student Practice
1. Multiply. $(7x+2)(7x-2)$	**2.** Multiply. $(8y+4)(8y-4)$
Use the rule for multiplying a sum and a difference.	
$(a+b)(a-b) = a^2 - b^2$	
Here, $a = 7x$ and $b = 2$.	
$(7x+2)(7x-2) = (7x)^2 - (2)^2$ $= 49x^2 - 4$	
The check is left to the student.	
3. Multiply. $(5x-8y)(5x+8y)$	**4.** Multiply. $(2x-9y)(2x+9y)$
Use the rule for multiplying a sum and a difference. Here, $a = 5x$ and $b = 8y$.	
$(5x-8y)(5x+8y) = (5x)^2 - (8y)^2$ $= 25x^2 - 64y^2$	

Vocabulary Answers: 1. vertical multiplication 2. square of a difference 3. FOIL

Example	Student Practice
5. Multiply. $(5y-2)^2$	**6.** Multiply. $(7x+y)^2$

5. Multiply. $(5y-2)^2$

Use a rule for a binomial squared.

$(a+b)^2 = a^2 + 2ab + b^2$

$(a-b)^2 = a^2 - 2ab + b^2$

Here, $a = 5y$ and $b = 2$.

$(5y-2)^2$

$= (5y)^2 - (2)(5y)(2) + (2)^2$

$= 25y^2 - 20y + 4$

6. Multiply. $(7x+y)^2$

---

7. Multiply vertically.

$(3x^3 + 2x^2 + x)(x^2 - 2x - 4)$

Place one polynomial over the other. Find the partial products and line them up under the original polynomials. Note that the answers for each partial product are placed so that like terms are underneath each other.

$$
\begin{array}{r}
3x^3 + 2x^2 + x \\
\underline{x^2 - 2x - 4} \\
-12x^3 - 8x^2 - 4x \\
-6x^4 - 4x^3 - 2x^2 \\
3x^5 - 2x^4 - x^3
\end{array}
$$

Find the sum of the three partial products.

$$
\begin{array}{r}
3x^3 + 2x^2 + x \\
\underline{x^2 - 2x - 4} \\
-12x^3 - 8x^2 - 4x \\
-6x^4 - 4x^3 - 2x^2 \\
\underline{3x^5 - 2x^4 - x^3 } \\
3x^5 - 4x^4 - 15x^3 - 10x^2 - 4x
\end{array}
$$

8. Multiply vertically.

$(x^2 + 3x - 6)(2x^2 - 4x - 5)$

Example	Student Practice
9. Multiply horizontally. $\left(x^2+3x+5\right)\left(x^2-2x-6\right)$ Use the distributive property repeatedly. $\left(x^2+3x+5\right)\left(x^2-2x-6\right)$ $=x^2\left(x^2-2x-6\right)+3x\left(x^2-2x-6\right)$ $\quad+5\left(x^2-2x-6\right)$ $=x^4-2x^3-6x^2+3x^3-6x^2-18x$ $\quad+5x^2-10x-30$ $=x^4+x^3-7x^2-28x-30$	**10.** Multiply horizontally. $\left(x^2-8x+10\right)\left(x^2+x+4\right)$
11. Multiply. $(2x-3)(x+2)(x+1)$ Multiply the first pair of binomials. Note that it does not matter which two binomials are multiplied first. $(2x-3)(x+2)=2x^2+4x-3x-6$ $\qquad\qquad\quad=2x^2+x-6$ Replace the first two factors with their resulting product. $\left(2x^2+x-6\right)(x+1)$ Multiply again and combine like terms. $\left(2x^2+x-6\right)(x+1)$ $=\left(2x^2+x-6\right)x+\left(2x^2+x-6\right)1$ $=2x^3+x^2-6x+2x^2+x-6$ $=2x^3+3x^2-5x-6$	**12.** Multiply. $(x-2)(x+3)(2x+4)$

135

Copyright © 2013 Pearson Education, Inc.

Extra Practice

1. Multiply. Use the special formula that applies. $(x+10)(x-10)$

2. Multiply. Use the special formula that applies. $(5a+3b)(5a-3b)$

3. Multiply. $(8x^2-2x+3)(3x+1)$

4. Multiply. $(x+2)(x-1)(x-5)$

Concept Check

Using the formula $(a+b)^2 = a^2 + 2ab + b^2$, explain how to multiply $(6x-9y)^2$.

Chapter 4 Exponents and Polynomials
4.6 Dividing Polynomials

Vocabulary
polynomial • subtraction • long division • descending order • binomial • monomial

1. To divide a polynomial by a _____, divide each term of the numerator by the denominator, then write the sum of the results.

2. _____ is a process used to divide polynomials when the divisor has two or more terms.

3. When dividing a polynomial by a binomial, you must first place the terms of each in _____, inserting a 0 for any missing terms.

4. When performing long division with polynomials, take great care on the_____ step when negative numbers are involved.

Example	**Student Practice**
1. Divide. $\dfrac{8y^6 - 8y^4 + 24y^2}{8y^2}$	**2.** Divide. $\dfrac{21x^5 + 7x^3 - 28x^2}{7x^2}$

Divide each term of the polynomial by the monomial.

$$\frac{8y^6 - 8y^4 + 24y^2}{8y^2} = \frac{8y^6}{8y^2} - \frac{8y^4}{8y^2} + \frac{24y^2}{8y^2}$$

Use the property $\dfrac{x^a}{x^b} = x^{a-b}$ to divide each term.

$$\frac{8y^6}{8y^2} - \frac{8y^4}{8y^2} + \frac{24y^2}{8y^2} = y^4 - y^2 + 3$$

Vocabulary Answers: 1. monomial 2. descending order 3. long division 4. subtraction

Example	Student Practice
3. Divide. $\left(x^3+5x^2+11x+4\right)\div\left(x+2\right)$	**4.** Divide. $\left(2x^3+4x^2+3x+6\right)\div\left(x+1\right)$

Notice the terms are already in descending order with no missing terms. Divide the first term of the polynomial by the first term of the binomial.

$$x+2\overline{\smash{)}x^3+5x^2+11x+4} \quad \overset{x^2}{}$$

Multiply x^2 by $x+2$ and subtract the result from the first two terms.

$$\begin{array}{r} x^2+3x \\ x+2\overline{\smash{)}x^3+5x^2+11x+4} \\ \underline{x^3+2x^2} \\ 3x^2+11x \end{array}$$

Continue this process until the degree of the remainder is less than the degree of the divisor.

$$\begin{array}{r} x^2+3x+5 \\ x+2\overline{\smash{)}x^3+5x^2+11x+4} \\ \underline{x^3+2x^2} \\ 3x^2+11x \\ \underline{3x^2+6x} \\ 5x+4 \\ \underline{5x+10} \\ -6 \end{array}$$

The remainder is written as the numerator of a fraction that has the binomial divisor as its denominator.

$$x^2+3x+5+\frac{-6}{x+2}$$

The check is left to the student.

Example	Student Practice

5. Divide. $\left(5x^3 - 24x^2 + 9\right) \div \left(5x + 1\right)$

Insert $0x$ into the polynomial to represent the missing x-term.

$$\left(5x^3 - 24x^2 + 0x + 9\right) \div \left(5x + 1\right)$$

Divide the first term of the polynomial, $5x^3$, by the first term of the binomial, $5x$.

$$\begin{array}{r} x^2 \\ 5x+1{\overline{\smash{\big)}\,5x^3 - 24x^2 + 0x + 9}} \\ \underline{5x^3 + x^2} \\ -25x^2 \end{array}$$

Divide $-25x^2$ by $5x$ and then divide $5x$ by $5x$. Be cautious with the negative signs.

$$\begin{array}{r} x^2 - 5x + 1 \\ 5x+1{\overline{\smash{\big)}\,5x^3 - 24x^2 + 0x + 9}} \\ \underline{5x^3 + x^2} \\ -25x^2 + 0x \\ \underline{-25x^2 - 5x} \\ 5x + 9 \\ \underline{5x + 1} \\ 8 \end{array}$$

Write the remainder as the numerator of a fraction that has the binomial divisor as its denominator.

The answer is $x^2 - 5x + 1 + \dfrac{8}{5x+1}$.

The check is left to the student.

6. Divide. $\left(8x^3 - 8x + 5\right) \div \left(2x + 1\right)$

Example	Student Practice
7. Divide and check.	**8.** Divide and check.
$\left(12x^3 - 11x^2 + 8x - 4\right) \div \left(3x - 2\right)$	$\left(8x^3 - 4x^2 + 6\right) \div \left(2x - 3\right)$

$$
\begin{array}{r}
4x^2 - x + 2 \\
3x - 2 \overline{\smash{\big)}\, 12x^3 - 11x^2 + 8x - 4} \\
\underline{12x^3 -\ 8x^2} \\
-3x^2 + 8x \\
\underline{-3x^2 + 2x} \\
6x - 4 \\
\underline{6x - 4} \\
0
\end{array}
$$

Check the answer.

$$\left(3x - 2\right)\left(4x^2 - x + 2\right) = 12x^3 - 11x^2 + 8x - 4$$

Extra Practice

1. Divide. $\dfrac{12a^7 - 4a^5 + 8a^3 - 2a^2}{2a^2}$

2. Divide. $\left(12y^4 - 18y^3 + 27y^2\right) \div 3y^2$

3. Divide and check. $\dfrac{3x^3 - 5x^2 + 7x - 5}{x - 1}$

Divide and check. $\dfrac{y^3 - 2y - 4}{y - 1}$

Concept Check

Explain how you would check if $x^2 + 2x + 8 + \dfrac{13}{x - 2}$ is the correct answer to the problem $\left(x^3 + 4x - 3\right) \div \left(x - 2\right)$. Perform the check. Does the answer check?

MATH COACH

Mastering the skills you need to do well on the test.

Watch the MATH COACH videos in MyMathLab® or on YouTube  while you work the problems below. These helpful hints will help you avoid making common errors on test problems.

Raising Monomials to a Power—Problem 8 Simplify $\dfrac{\left(3x^2\right)^3}{\left(6x\right)^2}$.

> **Helpful Hint:** Do the problem in three stages. First, use the power to a power rule to raise the numerator to the third power. Second, raise the denominator to the second power. Then divide the monomials using the rules of exponents. Be sure to simplify any fractions.

Did you use the power to a power rule to raise both 3^1 and x^2 to the third power in the numerator and both 6^1 and x^1 to the second power in the denominator?
Yes _____ No _____

If you answered No, stop and review the power to a power rule before completing these steps again.

If you answered No to either of these questions, go back and examine your work carefully before completing these steps again.

If you answered Problem 8 incorrectly, go back and rework the problem using these suggestions.

Did you remember to simplify the fraction $\dfrac{27}{36}$?

Yes _____ No _____

Finally, did you remember to use the quotient rule to subtract the exponents in the x terms?
Yes _____ No _____

Simplifying Monomials Involving Negative Exponents—Problem 11

Simplify and write with only positive exponents. $\dfrac{3x^{-3}y^2}{x^{-4}y^{-5}}$

> **Helpful Hint:** First, use the definition of a negative exponent to rewrite the expression using only positive exponents. Then use the rules for exponents to simplify the resulting expression.

Did you remove the negative exponents by rewriting the expression as $\dfrac{3x^4y^2y^5}{x^3}$? Yes _____ No _____

If you answered No, review the definition of negative exponents in Section 4.2 and complete this step again.

Did you use the quotient rule to simplify the x terms and the product rule to simplify the y terms? Yes _____ No _____

If you answered No, review the rules for exponents in Sections 4.1 and 4.2 and simplify the expression again.

Now go back and rework the problem using these suggestions.

141

Multiplying Three Binomials—Problem 20   Multiply $(3x+2)(2x+1)(x-3)$.

Helpful Hint: A good approach is to start by multiplying the first two binomials. Then multiply that result by the third binomial. Be careful to avoid sign errors when multiplying, and be careful to write down the correct exponent for each term.

Did you use the FOIL method to multiply the first two binomials and obtain $6x^2+7x+2$? Yes _____ No _____

If you answered No, stop and complete this step.

Did you multiply the result above by $(x-3)$?
Yes _____ No _____

Did you multiply each term of $6x^2+7x+2$ by x?
Yes _____ No _____

Did you multiply each term of $6x^2+7x+2$ by -3?
Yes _____ No _____

If you answered No to any of these questions, go back and examine each step of the multiplication carefully.

Be sure to write the correct exponent each time that you multiply. Be careful to avoid sign errors when multiplying by -3. Then combine like terms before writing your final answer.

If you answered Problem 20 incorrectly, go back and rework the problem using these suggestions.

Dividing a Polynomial by a Binomial—Problem 27   Divide $(2x^3-6x-36)\div(x-3)$.

Helpful Hint: Review the procedure for dividing a polynomial by a binomial in Section 4.6. Make sure you understand each step. Be sure you understand where the expression $0x^2$ came from in the dividend. Be careful with subtraction. Write out the subtraction steps to avoid sign errors.

Did you write the division problem in the form
$x-3\overline{)2x^3+0x^2-6x-36}$? Yes _____ No _____
If you answered No, remember that every power must be represented. We must use $0x^2$ as a placeholder so that we can perform our division.

When you carried out the first step of division, did you obtain $2x^2$ as the first part of your answer?
Yes _____ No _____

When you multiplied $x-3$ by $2x^2$, and then subtracted, did you get the result $6x^2$? Yes _____ No _____

If you answered No to these questions, stop and examine your first division step carefully. Make sure that you subtracted carefully too. Remember to write out the subtraction steps: $0x^2-\left(-6x^2\right)=6x^2$.

Next, did you bring down $-6x$ from the dividend to obtain $6x^2-6x$?
Yes _____ No _____

If you answered No, go back and look at the dividend again and see how to obtain this result.

Now go back and rework the problem using these suggestions.

Name: _____   Date: _____

Instructor: _____   Section: _____

Chapter 5 Factoring
5.1 Removing a Common Factor

Vocabulary
factor   •   to factor   •   common factor   •   greatest common factor

1. When two or more numbers, variables, or algebraic expressions are multiplied, each is called a _____.

2. A _____ is a factor that both terms have in common.

3. When you are asked _____ a number or an algebraic expression, you are being asked, "What factors, when multiplied, will give that number or expression?"

4. When we factor, we begin by looking for the _____.

Example	**Student Practice**
1. Factor. (a) $3x - 6y$	**2.** Factor. (a) $5y + 15z$
Begin by looking for a common factor, a factor that both terms have in common. Then rewrite the expression as a product. $3x - 6y = 3(x - 2y)$ This is true because $3(x - 2y) = 3x - 6y.$	
(b) $9x + 2xy$ $9x + 2xy = x(9 + 2y)$ This is true because $x(9 + 2y) = 9x + 2xy.$ Notice that factoring is using the distributive property in reverse.	(b) $12y - 5yz$

Vocabulary Answers: 1. factor 2. common factor 3. to factor 4. greatest common factor

Example	Student Practice
3. Factor $24xy + 12x^2 + 36x^3$. Remember to remove the greatest common factor.	**4.** Factor $44y^3 + 55y^2 - 11xy$. Remember to remove the greatest common factor.

3. Factor $24xy + 12x^2 + 36x^3$. Remember to remove the greatest common factor.

We start by finding the greatest common factor of 24, 12, and 36.

You may want to factor each number, or you may notice that 12 is a common factor. 12 is the greatest numerical common factor.

Notice also that x is a factor of each term. Thus, $12x$ is the greatest common factor.

$$24xy + 12x^2 + 36x^3 = 12x\left(2y + x + 3x^2\right)$$

4. Factor $44y^3 + 55y^2 - 11xy$. Remember to remove the greatest common factor.

---

5. Factor.

(a) $12x^2 - 18y^2$

Note that the largest integer that is common to both terms is 6 (not 3 or 2).

$$12x^2 - 18y^2 = 6\left(2x^2 - 3y^2\right)$$

(b) $x^2y^2 + 3xy^2 + y^3$

Although y is common to all of the terms, we factor out y^2 since 2 is the largest exponent of y that is common to all terms.

We do not factor out x, since x is not common to all of the terms.

$$x^2y^2 + 3xy^2 + y^3 = y^2\left(x^2 + 3x + y\right)$$

6. Factor.

(a) $21m^2 - 28n^2$

(b) $m^3n^2 + 9m^2n^2 + 3m^4$

Example	Student Practice
7. Factor. $8x^3y + 16x^2y^2 - 24x^3y^3$	**8.** Factor. $27xy^3 - 36x^2y^2 - 9x^3y^3$

We see that 8 is the largest integer that will divide evenly into the three numerical coefficients. We can factor x^2 out of each term. We can also factor y out of each term.

$$8x^3y + 16x^2y^2 - 24x^3y^3$$
$$= 8x^2y(x + 2y - 3xy^2)$$

Check.

$$8x^2y(x + 2y - 3xy^2)$$
$$= 8x^3y + 16x^2y^2 - 24x^3y^3$$

9. Factor. $9a^3b^2 + 9a^2b^2$	**10.** Factor. $11m^4n^2 + 11m^3n^2$

We observe that both terms contain a common factor of 9. We can also factor a^2 and b^2 out of each term. Thus, the greatest common factor is $9a^2b^2$.

$$9a^3b^2 + 9a^2b^2 = 9a^2b^2(a + 1)$$

11. Factor. $7x^2(2x - 3y) - (2x - 3y)$	**12.** Factor. $5y^2(3x + 4y) - (3x + 4y)$

The common factor of the terms is $(2x - 3y)$. What happens when we factor out $(2x - 3y)$? What are we left with in the second term?
Recall that $(2x - 3y) = 1(2x - 3y)$.

$$7x^2(2x - 3y) - (2x - 3y)$$
$$= 7x^2(2x - 3y) - \mathbf{1}(2x - 3y)$$
$$= (2x - 3y)(7x^2 - 1)$$

145

Example	Student Practice
13. A computer programmer is writing a program to find the total area of 4 circles. She uses the formula $A = \pi r^2$. The radii of the circles are a, b, c, and d, respectively. She wants the final answer to be in factored form with the value of π occurring only once, in order to minimize the rounding error. Write the total area of the 4 circles with a formula that has π occurring only once.	**14.** Find the total area of 3 circles using the formula $A = \pi r^2$. The radii of the circles are m, $2n$, and $3z$, respectively. Write the total area of the 3 circles with a formula that has π occurring only once.

For each circle, $A = \pi r^2$, where $r = a$, b, c, or d. We add the area of each of the 4 circles.

The total area is $\pi a^2 + \pi b^2 + \pi c^2 + \pi d^2$.

In factored form the total area =
$\pi\left(a^2 + b^2 + c^2 + d^2\right)$.

Extra Practice

1. Remove the largest possible common factor. Check your answer by multiplication. $3x^3 + 12x^2 - 21x$

2. Remove the largest possible common factor. Check your answer by multiplication. $60x^2 y + 18xy - 24x$

3. Remove the largest possible common factor. Check your answer by multiplication. $8a(x+3y) - b(x+3y)$

4. Remove the largest possible common factor. Check your answer by multiplication. $5a(4x-3) - (4x-3)$

Concept Check
Explain how you would remove the greatest common factor from the following polynomial. $36a^3b^2 - 72a^2b^3$

Chapter 5 Factoring
5.2 Factoring by Grouping

Vocabulary
factoring by grouping • common factor • commutative property • FOIL

1. Sometimes you will need to factor out a negative _____ from the second two terms to obtain two terms that contain the same parenthetical expression.

2. A procedure used to factor a four-term polynomial is called _____.

3. Rearrange the order using the _____ of addition.

4. To check, we multiply the two binomials using the _____ procedure.

Example	Student Practice
1. Factor. $x(x-3)+2(x-3)$ Observe each term: $\underbrace{x(x-3)}_{\substack{\text{first}\\\text{term}}}+\underbrace{2(x-3)}_{\substack{\text{second}\\\text{term}}}$ The common factor of the first and second terms is the quantity $(x-3)$. $x(x-3)+2(x-3)=(x-3)(x+2)$	**2.** Factor. $z(2z+5)-4(2z+5)$
3. Factor. $2x^2+3x+6x+9$ Factor out a common factor of x from the first two terms. Factor out a common factor of 3 from the second two terms. $2x^2+3x+6x+9 = x(2x+3)+3(2x+3)$ The expression in parentheses is now a common factor of the terms. $x(2x+3)+3(2x+3)=(2x+3)(x+3)$	**4.** Factor. $12x^2+4x+15x+5$

Vocabulary Answers: 1. common factor 2. factoring by grouping 3. commutative property 4. FOIL

Example	Student Practice
5. Factor. $4x + 8y + ax + 2ay$	**6.** Factor. $7y + 21z + xy + 3xz$

5. Factor. $4x + 8y + ax + 2ay$

Factor out a common factor of 4 from the first two terms. Factor out a common factor of a from the second two terms.

$4x + 8y + ax + 2ay$

$= 4(x + 2y) + a(x + 2y)$

The common factor of the terms is the expression in parentheses, $x + 2y$.

$4(x + 2y) + a(x + 2y) = (x + 2y)(4 + a)$

---

7. Factor. $bx + 4y + 4b + xy$

Rearrange the terms so that the first two terms have a common factor.

$bx + 4y + 4b + xy = bx + 4b + xy + 4y$

Factor out the common factor b from the first two terms and the common factor y from the second two terms.

$bx + 4b + xy + 4y = b(x + 4) + y(x + 4)$

$\qquad\qquad\qquad = (x + 4)(b + y)$

8. Factor. $nz + 6y + 6z + ny$

---

9. Factor. $2x^2 + 5x - 4x - 10$

Factor out the common factor x from the first two terms and the common factor -2 from the second two terms.

$2x^2 + 5x - 4x - 10$

$= x(2x + 5) - 4x - 10$

$= x(2x + 5) - 2(2x + 5)$

$= (2x + 5)(x - 2)$

10. Factor. $2x^2 + 3x - 6x - 9$

148

Example	Student Practice
11. Factor. $2ax - a - 2bx + b$	**12.** Factor. $5nz - n - 5mz + m$

11. (continued)

Factor out the common factor a from the first two terms and the common factor $-b$ from the second two terms.

$$2ax - a - 2bx + b = a(2x - 1) - b(2x - 1)$$

Since the two resulting terms contain the same parenthetical expression, we can complete the factoring.

$$a(2x - 1) - b(2x - 1) = (2x - 1)(a - b)$$

---

13. Factor and check your answer.
$$8ad + 21bc - 6bd - 28ac$$

Use the commutative property of addition to rearrange the order so the first two terms have a common factor.

$$8ad + 21bc - 6bd - 28ac$$
$$= 8ad - 6bd - 28ac + 21bc$$

Factor out the common factor $2d$ from the first two terms and the common factor $-7c$ from the last two terms.

$$= 2d(4a - 3b) - 7c(4a - 3b)$$

Factor out the common factor $(4a - 3b)$.

$$= (4a - 3b)(2d - 7c)$$

To check, we multiply the two binomials using the FOIL procedure.

$$(4a - 3b)(2d - 7c)$$
$$= 8ad - 28ac - 6bd + 21bc$$
$$= 8ad + 21bc - 6bd - 28ac$$

14. Factor and check your answer.
$$9wz + 35xy - 15xz - 21wy$$

Extra Practice

1. Factor by grouping. Check you answer.

    $x(x+1) - 2(x+1)$

2. Factor by grouping. Check you answer.

    $x^2 - 2x + 5x - 10$

3. Factor by grouping. Check you answer.

    $6x^2 + 15x - 4x - 10$

4. Factor by grouping. Check you answer.

    $12ax + 15ay + 8xb + 10by$

Concept Check

Explain how you would factor the following polynomial.

$10ax + b^2 + 2bx + 5ab$

Name: _____   Date: _____

Instructor: _____   Section: _____

Chapter 5 Factoring

5.3 Factoring Trinomials of the Form $x^2 + bx + c$

Vocabulary

first terms   •   outer and inner terms   •   second terms   •   last terms

1. When multiplying an expression of the form $(x+m)(x+n)$, the product of the
_____ in the factors produces the first term of the polynomial.

2. When multiplying an expression of the form $(x+m)(x+n)$, the sum of the
_____ in the factors gives the coefficient of the middle term of the polynomial.

3. When multiplying an expression of the form $(x+m)(x+n)$, the sum of the products of
the _____ in the factors produces the middle term of the polynomial.

4. When multiplying an expression of the form $(x+m)(x+n)$, the product of the
_____ of the factors gives the last term of the polynomial.

Example	Student Practice
1. Factor. $x^2 + 7x + 12$ The answer will be of the form $(x+m)(x+n)$. We want to find the two numbers, m and n, that you can multiply to get 12 and add to get 7. The numbers are 3 and 4. $x^2 + 7x + 12 = (x+3)(x+4)$	**2.** Factor. $x^2 + 9x + 18$
3. Factor. $x^2 + 12x + 20$ We want two numbers that have a product of 20 and a sum of 12. The numbers are 10 and 2. $x^2 + 12x + 20 = (x+10)(x+2)$	**4.** Factor. $x^2 + 14x + 33$

Vocabulary Answers: 1. first terms 2. second terms 3. outer and inner terms 4. last terms

Example	Student Practice
5. Factor. $x^2 - 8x + 15$	**6.** Factor. $x^2 - 11x + 28$

5. Factor. $x^2 - 8x + 15$

We want two numbers that have a product of $+15$ and a sum of -8. They must be negative numbers since the sign of the middle term is negative and the sign of the last term is positive.

The sum $-5 + (-3)$ is -8 and the product $(-5)(-3)$ is $+15$.

$$x^2 - 8x + 15 = (x - 5)(x - 3)$$

Multiply using FOIL to check.

6. Factor. $x^2 - 11x + 28$

---

7. Factor. $x^2 - 3x - 10$

We want two numbers whose product is -10 and whose sum is -3. The two numbers are -5 and $+2$.

$$x^2 - 3x - 10 = (x - 5)(x + 2)$$

8. Factor. $x^2 - 2x - 24$

---

9. Factor and check your answer.
$$y^2 + 10y - 24$$

The two numbers whose product is -24 and whose sum is $+10$ are $+12$ and -2.

$$y^2 + 10y - 24 = (y + 12)(y - 2)$$

It is very easy to make a sign error in these problems. Make sure that you mentally check your answer by FOIL to obtain the original expression.

Check.

$$(y + 12)(y - 2) = y^2 - 2y + 12y - 24$$
$$= y^2 + 10y - 24$$

10. Factor and check your answer.
$$z^2 + 12z - 28$$

152

Example	Student Practice
11. Factor. $y^4 - 2y^2 - 35$	**12.** Factor. $x^4 - 3x^2 - 10$

11. Factor. $y^4 - 2y^2 - 35$

Notice that $y^4 = (y^2)(y^2)$. This will be the first term in each set of parentheses.

$$(y^2 \quad)(y^2 \quad)$$

The last term of the polynomial is negative. Thus, the signs of m and n will be different.

$$(y^2 + \quad)(y^2 - \quad)$$

Now think of factors of 35 whose difference is 2.

$$(y^2 + 5)(y^2 - 7)$$

Multiply using FOIL to check.

---

13. Factor. $3x^2 + 9x - 162$

First factor out the common factor 3 from each term of the polynomial.

$$3x^2 + 9x - 162 = 3(x^2 + 3x - 54)$$

Then factor the remaining polynomial.

$$3(x^2 + 3x - 54) = 3(x - 6)(x + 9)$$

The final answer is $3(x - 6)(x + 9)$. Be sure to include the 3.

Check.

$$3(x - 6)(x + 9) = 3(x^2 + 3x - 54)$$
$$= 3x^2 + 9x - 162$$

14. Factor. $4x^2 - 16x - 84$

Example	**Student Practice**

15. Find a polynomial in factored form for the shaded area in the figure.

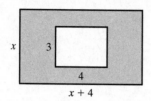

16. Find a polynomial in factored form for the shaded area in the figure.

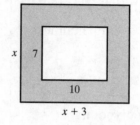

To obtain the shaded area, we find the area of the larger rectangle and subtract from it the area of the smaller rectangle. Thus, we have the following:

$$\text{shaded area} = x(x+4)-(4)(3)$$
$$= x^2 +4x-12$$

Now we factor this polynomial to obtain the shaded area $=(x+6)(x-2)$.

Extra Practice

1. Factor. $x^2 +6x+8$

2. Factor. $a^2 -7a+12$

3. Factor. $x^2 -x-12$

4. Factor. $2x^2 +18x+28$

Concept Check

Explain how you would completely factor $4x^2 -4x-120$.

Chapter 5 Factoring
5.4 Factoring Trinomials of the Form $ax^2 + bx + c$

Vocabulary
trial-and-error method   •   grouping method   •   multiplying   •   greatest common factor

1. The _____ requires listing the possible factoring combinations and computing the middle terms by the FOIL method.

2. Always check by _____ the factors to see if the original trinomial is obtained.

3. First factor out the _____ if any, and then factor the trinomial.

4. The _____ for factoring trinomials of the form $ax^2 + bx + c$ requires writing the polynomial with four terms and then factoring the polynomial by grouping.

Example	Student Practice
1. Factor. $2x^2 + 5x + 3$	**2.** Factor. $2x^2 + 13x + 20$

Example

1. Factor. $2x^2 + 5x + 3$

In order for the coefficient of the x^2- term of the polynomial to be 2, the coefficients of the x-terms in the factors must be 2 and 1. Thus,

$$2x^2 + 5x + 3 = (2x\quad)(x\quad).$$

In order for the last term of the polynomial to be 3, the constants in the factors must be 3 and 1. Since all signs in the polynomial are positive, each factor in parentheses will be positive. We have two possibilities. We check them by multiplying.

$$(2x+1)(x+3) = 2x^2 + 7x + 3$$
$$(2x+3)(x+1) = 2x^2 + 5x + 3$$

The correct answer is $(2x+3)(x+1)$.

Vocabulary Answers: 1. trial-and-error method 2. multiplying 3. greatest common factor 4. grouping method

Example	Student Practice
3. Factor. $4x^2 - 13x + 3$	**4.** Factor. $3x^2 - 4x + 1$

The different factorizations of 4 are $(2)(2)$ and $(1)(4)$. The factorization of 3 is $(1)(3)$. Let us list the possible factoring combinations and compute the middle term by the FOIL method. Note that the signs of the constants in both factors will be negative.

Possible Factors	Middle Term
$(2x-3)(2x-1)$	$-8x$
$(4x-3)(x-1)$	$-7x$
$(4x-1)(x-3)$	$-13x$

The correct middle term is $-13x$. The correct answer is $(4x-1)(x-3)$.

---

5. Factor. $3x^2 - 2x - 8$	**6.** Factor. $4x^2 - 5x - 21$

The factorization of 3 is $(1)(3)$. The factorizations of 8 are $(8)(1)$ and $(4)(2)$. Let us list only one-half of the possibilities.

Possible Factors	Middle Term
$(x+8)(3x-1)$	$+23x$
$(x+1)(3x-8)$	$-5x$
$(x+4)(3x-2)$	$+10x$
$(x+2)(3x-4)$	$+2x$

The middle term $+2x$ is only incorrect because the sign is wrong. So we just reverse the signs of the constraints.

The correct answer is $(x-2)(3x+4)$.

Example	Student Practice

7. Factor by grouping. $2x^2 + 5x + 3$

First find the grouping number. The grouping number is $(2)(3) = 6$. The factors of 6 are $6 \cdot 1$ and $3 \cdot 2$. We choose the numbers 3 and 2 because their product is 6 and their sum is 5. Next, write $5x$ as the sum $2x + 3x$. Then, factor by grouping.

$$2x^2 + 5x + 3 = 2x^2 + 2x + 3x + 3$$
$$= 2x(x+1) + 3(x+1)$$
$$= (x+1)(2x+3)$$

Multiply to check.

$$(x+1)(2x+3) = 2x^2 + 3x + 2x + 3$$
$$= 2x^2 + 5x + 3$$

8. Factor by grouping. $2x^2 + 5x + 3$

9. Factor by grouping. $3x^2 - 2x - 8$

First find the grouping number. The grouping number is $(3)(-8) = -24$. Now we want two numbers whose product is -24 and whose sum is -2. They are -6 and 4. So we write $-2x$ as the sum $-6x + 4x$. Then, we factor by grouping.

$$3x^2 - 6x + 4x - 8 = 3x(x-2) + 4(x-2)$$
$$= (x-2)(3x+4)$$

10. Factor by grouping. $5x^2 - 13x - 6$

Example	Student Practice
11. Factor. $9x^2 + 3x - 30$	**12.** Factor. $8x^2 + 22x - 6$

11. Factor. $9x^2 + 3x - 30$

We first factor out the common factor 3 from each term of the trinomial.

$$9x^2 + 3x - 30 = 3\left(3x^2 + 1x - 10\right)$$

We then factor the trinomial by the grouping method or by the trial-and-error method.

$$3\left(3x^2 + 1x - 10\right) = 3(3x - 5)(x + 2)$$

---

13. Factor. $32x^2 - 40x + 12$ **14.** Factor. $30x^2 - 65x + 30$

We first factor out the greatest common factor 4 from each term of the trinomial.

$$32x^2 - 40x + 12 = 4\left(8x^2 - 10x + 3\right)$$

We then factor the trinomial by the grouping method or by the trial-and-error method.

$$4\left(8x^2 - 10x + 3\right) = 4(2x - 1)(4x - 3)$$

Extra Practice

1. Factor by the trial-and-error method. Check your answer by using FOIL.
$4x^2 - 12x + 5$

2. Factor by the grouping method. Check your answer by using FOIL. $7x^2 - 4x - 3$

3. Factor by any method. $5x^2 + 43x - 18$

4. Factor by first factoring out the greatest common factor. $15y^2 + 51y - 36$

Concept Check

Explain how you would factor $10x^3 + 18x^2y - 4xy^2$.

Name: _____   Date: _____
Instructor: _____   Section: _____

Chapter 5 Factoring
5.5 Special Cases of Factoring

Vocabulary
difference of two squares • negative • perfect-square trinomials • greatest common factor

1. The difference of two squares formula only works if the last term is _____.

2. A _____ is a trinomial where the first and last terms are perfect squares and the middle term is twice the product of the values whose squares are the first and last terms.

3. The _____ can be factored into the sum and difference of those values that were squared.

4. For some polynomials, we first need to factor out the _____.

Example	**Student Practice**
1. Factor. $9x^2 - 1$	**2.** Factor. $36x^2 - 1$
We see that the polynomial is in the form of the difference of two squares. $9x^2$ is a square and 1 is a square. $9x^2 = (3x)^2$ and $1 = (1)^2$. So using the formula we can write the following.	
$9x^2 - 1 = (3x+1)(3x-1)$	
3. Factor. $25x^2 - 16$	**4.** Factor. $64x^2 - 9$
$25x^2 = (5x)^2$ and $16 = (4)^2$. Again we use the formula for the difference of two squares.	
$25x^2 - 16 = (5x+4)(5x-4)$	

Vocabulary Answers: 1. negative 2. perfect-square trinomials 3. difference of two squares 4. greatest common factor

Example	Student Practice
5. Factor. $81x^4 - 1$	**6.** Factor. $16x^4 - 1$

5. Factor. $81x^4 - 1$

Because $81x^4 = \left(9x^2\right)^2$ and $1 = \left(1\right)^2$, we see that

$$81x^4 - 1 = \left(9x^2 + 1\right)\left(9x^2 - 1\right).$$

The factoring is not complete. We can factor $9x^2 - 1$, because
$9x^2 - 1 = \left(3x + 1\right)\left(3x - 1\right)$.

$$81x^4 - 1 = \left(9x^2 + 1\right)\left(3x + 1\right)\left(3x - 1\right)$$

---

7. Factor. $x^2 + 6x + 9$

This is a perfect-square trinomial. The first and last terms are perfect squares because $x^2 = \left(x\right)^2$ and $9 = \left(3\right)^2$. The middle term, $6x$, is twice the product of 3 and x.

Since $x^2 + 6x + 9$ is a perfect-square trinomial we can use the formula

$$a^2 + 2ab + b^2 = \left(a + b\right)^2$$

with $a = x$ and $b = 3$. So we have

$$x^2 + 6x + 9 = \left(x + 3\right)^2.$$

8. Factor. $x^2 + 10x + 25$

Example	Student Practice
9. Factor.	**10.** Factor.

9. Factor.

(a) $49x^2 + 42xy + 9y^2$

This is a perfect square trinomial, because $49x^2 = (7x)^2$, $9y^2 = (3y)^2$, and $42xy = 2(7x \cdot 3y)$.

$$49x^2 + 42xy + 9y^2 = (7x + 3y)^2$$

(b) $36x^4 - 12x^2 + 1$

This is a perfect square trinomial, because $36x^4 = (6x^2)^2$, $1 = (1)^2$, and $12x^2 = 2(6x^2 \cdot 1)$.

$$36x^4 - 12x^2 + 1 = (6x^2 - 1)^2$$

10. Factor.

(a) $64x^2 + 80xy + 25y^2$

(b) $49x^4 - 14x^2 + 1$

11. Factor. $49x^2 + 35x + 4$

This is not a perfect-square trinomial! Although the first and last terms are perfect squares since $(7x)^2 = 49x^2$ and $(2)^2 = 4$, the middle term, $35x$, is not double the product of 2 and $7x$. $35x \neq 28x$! So we must factor by trial and error or by grouping to obtain

$$49x^2 + 35x + 4 = (7x + 4)(7x + 1).$$

12. Factor. $36x^2 + 25x + 4$

Example	Student Practice
13. Factor. $12x^2 - 48$	**14.** Factor. $3x^2 - 75$

We see that the greatest common factor is 12. First we factor out 12. Then we use the difference-of-two-squares formula, $a^2 - b^2 = (a+b)(a-b)$.

$$12x^2 - 48 = 12(x^2 - 4)$$
$$= 12(x+2)(x-2)$$

Example	Student Practice
15. Factor. $24x^2 - 72x + 54$	**16.** Factor. $48x^2 - 72x + 27$

First we factor out the greatest common factor, 6. Then we use the perfect-square-trinomial formula,
$a^2 - 2ab + b^2 = (a-b)^2$.

$$24x^2 - 72x + 54 = 6(4x^2 - 12x + 9)$$
$$= 6(2x-3)^2$$

Extra Practice

1. Factor using the difference-of-two-squares formula. $25a^2 - 36b^2$

2. Factor by using the perfect-square trinomial formula. $36a^2 + 60ab + 25b^2$

3. Factor by using the difference-of-two-squares. $x^4 - 100$

4. Factor by using the perfect-square trinomial formula. $9x^4 - 30x^2 + 25$

Concept Check

Explain how to factor the polynomial $24x^2 + 120x + 150$.

Name: _____   Date: _____

Instructor: _____   Section: _____

Chapter 5 Factoring
5.6 A Brief Review of Factoring

Vocabulary
perfect-square trinomial   •   difference of two squares   •   factor by grouping   •   prime

trinomial of the form $x^2 + bx + c$ •   trinomial of the form $ax^2 + bx + c$ •   common factor

1. Some polynomials have a _____ consisting of a number, a variable, or both.

2. When we _____, we rearrange the order if the first two terms do not have a common factor.

3. If we cannot factor a polynomial by elementary methods, we will identify it as a _____ polynomial.

4. In a _____ there are three terms and the first and last terms are perfect squares.

Example	Student Practice
1. Factor.	**2.** Factor.
(a) $25x^3 - 10x^2 + x$	**(a)** $9y^3 + 6y^2 + y$
Factor out the common factor x. The other factor is a perfect-square trinomial.	
$25x^3 - 10x^2 + x = x\left(25x^2 - 10x + 1\right)$ $= x(5x-1)^2$	
(b) $20x^2 y^2 - 45 y^2$	**(b)** $-4x^3 + 28x^2 + 32x$
Factor out the common factor $5y^2$. The other factor is a difference of squares.	
$20x^2 y^2 - 45 y^2 = 5y^2\left(4x^2 - 9\right)$ $= 5y^2(2x+3)(2x-3)$	

Vocabulary Answers: 1. common factor 2. factor by grouping 3. prime 4. perfect-square trinomial

Example	Student Practice
3. Factor. $ax^2 - 9a + 2x^2 - 18$	**4.** Factor. $bx^2 - 16b - 3x^2 + 48$

3. Factor. $ax^2 - 9a + 2x^2 - 18$

We factor by grouping since there are four terms. Factor out the common factor a from the first two terms and 2 from the second two terms.

$ax^2 - 9a + 2x^2 - 18$

$= a(x^2 - 9) + 2(x^2 - 9)$

Factor out the common factor $(x^2 - 9)$.

$a(x^2 - 9) + 2(x^2 - 9) = (a + 2)(x^2 - 9)$

Factor $x^2 - 9$ using the difference-of-two squares formula.

$(a + 2)(x^2 - 9) = (a + 2)(x - 3)(x + 3)$

---

5. Factor, if possible.

(a) $x^2 + 6x + 12$

The factors of 12 are

$(1)(12)$ or $(2)(6)$ or $(3)(4)$.

None of these pairs add up to 6, the coefficient of the middle term. Thus, the problem cannot be factored by the methods in this chapter.

(b) $25x^2 + 4$

We have a formula to factor the difference of two squares. There is no way to factor the sum of two squares. That is, $a^2 + b^2$ cannot be factored.

6. Factor, if possible.

(a) $x^2 + 5x + 17$

(b) $x^2 - x + 21$

Extra Practice

1. Factor, if possible. Be sure to factor completely. Factor out the greatest common factor first, if one exists.

   $x^2 + 9$

2. Factor, if possible. Be sure to factor completely. Factor out the greatest common factor first, if one exists.

   $63x - 7x^3$

3. Factor, if possible. Be sure to factor completely. Factor out the greatest common factor first, if one exists.

   $-x^3 + 12x^2 + 45x$

4. Factor, if possible. Be sure to factor completely. Factor out the greatest common factor first, if one exists.

   $x^2 + 2x + 3x + 6$

Concept Check

Explain how to completely factor $2x^2 + 6xw - 5x - 15w$.

166

Chapter 5 Factoring
5.7 Solving Quadratic Equations by Factoring

Vocabulary
quadratic equation　　•　　standard form　　•　　real roots　　•　　zero factor property

1. Many quadratic equations have two real number solutions, also called _____.

2. The _____ states that if $a \cdot b = 0$, then $a = 0$ or $b = 0$.

3. A _____ is a polynomial equation in one variable that contains a variable term of degree 2 and no terms of higher degree.

Example	Student Practice
1. Solve the equation to find the two roots. $$2x^2 + 13x - 7 = 0$$ The equation is in standard form. Factor. Then set each factor equal to 0 and solve the equations to find the two roots. $$2x^2 + 13x - 7 = 0$$ $$(2x - 1)(x + 7) = 0$$ $$2x - 1 = 0 \quad x + 7 = 0$$ $$x = \frac{1}{2} \qquad x = -7$$ Check. If $x = \frac{1}{2}$ and if $x = -7$ then we have the following. $$2\left(\frac{1}{2}\right)^2 + 13\left(\frac{1}{2}\right) - 7 = \frac{1}{2} + \frac{13}{2} - \frac{14}{2} = 0$$ $$2(-7)^2 + 13(-7) - 7 = 98 - 91 - 7 = 0$$ Thus $\frac{1}{2}$ and -7 are both roots of the equation $2x^2 + 13x - 7 = 0$.	**2.** Solve the equation to find the two roots. $$3x^2 + 8x - 3 = 0$$

Vocabulary Answers: 1. real roots 2. zero factor property 3. quadratic equation

Example	Student Practice
3. Solve the equation to find the two roots. $7x^2 - 3x = 0$	**4.** Solve the equation to find the two roots. $5x^2 - 4x = 0$

3. The equation is in standard form. Here $c = 0$. Factor out the common factor. Then set each factor equal to 0 by the zero factor property. Solve the equations to find the two roots.

$$7x^2 - 3x = 0$$
$$x(7x - 3) = 0$$
$$x = 0 \quad 7x - 3 = 0$$
$$7x = 3$$
$$x = \frac{3}{7}$$

The two roots are 0 and $\frac{3}{7}$.

Check. Verify that 0 and $\frac{3}{7}$ are the roots of $7x^2 - 3x = 0$.

---

5. Solve. $x^2 = 12 - x$	**6.** Solve. $x^2 = 63 - 2x$

5. The equation is not in standard form. Add x and -12 to both sides of the equation so that the left side is equal to zero. Then factor and set each factor equal to 0. Solve the equations for x.

$$x^2 = 12 - x$$
$$x^2 + x - 12 = 0$$
$$(x - 3)(x + 4) = 0$$
$$x - 3 = 0 \quad x + 4 = 0$$
$$x = 3 \quad x = -4$$

The check is left to the student.

Example	Student Practice

7. Carlos lives in Mexico City. He has a rectangular brick walkway in front of his house. The length of the walkway is 3 meters longer than twice the width. The area of the walkway is 44 square meters. Find the length and width of the rectangular walkway.

Let w = the width in meters.

Then $2w + 3$ = the length in meters.

Next, write an equation.

$$\text{area} = (\text{width})(\text{length})$$
$$44 = w(2w + 3)$$

Now, solve and state the answer.

$$44 = w(2w + 3)$$
$$44 = 2w^2 + 3w$$
$$0 = 2w^2 + 3w - 44$$
$$0 = (2w + 11)(w - 4)$$
$$2w + 11 = 0 \qquad w - 4 = 0$$
$$w = -5\frac{1}{2} \qquad w = 4$$

Since it would not make sense to have a rectangle with a negative number as a width, $-5\frac{1}{2}$ is not a valid solution.

Since $w = 4$, the width of the walkway is 4 meters. The length is $2w + 3$, so we have $2(4) + 3 = 8 + 3 = 11$. Thus, the length of the walkway is 11 meters.

The check is left to the student.

8. The length of a rectangle is 31 inches shorter than triple the width. The rectangle has an area of 60 square inches. Find the length and width of the rectangle.

169

Example	Student Practice
9. A tennis ball is thrown upward with an initial velocity of 8 meters/second. Suppose the initial height above the ground is 4 meters. At what time t will the ball hit the ground?	**10.** A baseball is thrown upward with an initial velocity of 9 meters/second. Suppose the initial height above the ground is 18 meters. At what time t will the ball hit the ground?

In this case $S = 0$ since the ball will hit the ground. The initial upward velocity is $v = 8$ meters/second. The initial height is 4 meters, so $h = 4$.

$$S = -5t^2 + vt + h$$
$$0 = -5t^2 + 8t + 4$$
$$5t^2 - 8t - 4 = 0$$
$$(5t + 2)(t - 2) = 0$$
$$5t + 2 = 0 \qquad t - 2 = 0$$
$$t = -\frac{2}{5} \qquad t = 2$$

We want a positive time for t in seconds; thus we do not use $t = -\frac{2}{5}$.

Therefore, the ball will strike the ground 2 seconds after it is thrown.

The check is left to the student.

Extra Practice

1. Solve for the roots of the quadratic equation. Check your answer.
$2x^2 - 5x - 3 = 0$

2. Solve for the roots of the quadratic equation. Check your answer.
$3x^2 = 27$

3. Solve for the roots of the quadratic equation. Check your answer.
$x^2 - 35 = 2x$

4. Solve for the roots of the quadratic equation. Check your answer.
$x^2 - 7x = -12$

Concept Check

Explain how you would solve the following problem: A rectangle has an area of 65 square feet. The length of the rectangle is 3 feet longer than double the width. Find the length and the width of the rectangle.

MATH COACH

Mastering the skills you need to do well on the test.

Watch the MATH COACH videos in MyMathLab® or on YouTube
while you work the problems below. These helpful hints will
help you avoid making common errors on test problems.

Factoring the Difference of Two Squares—Problem 2

Factor completely. $16x^2 - 81$

> **Helpful Hint:** It is important to learn the difference-of-two-squares
> formula: $a^2 - b^2 = (a+b)(a-b)$. Remember that the numerical values in
> both terms will be perfect squares. The first ten perfect squares are 1, 4, 9,
> 16, 25, 36, 49, 64, 81, and 100.

Did you remember that $(4x)^2 = 16x^2$? Yes _____ No _____

Did you remember that $9^2 = 81$? Yes _____ No _____

If you answered No to these questions, stop and review the
list of the first ten perfect squares. Consider that $4^2 = 16$ and
$x \cdot x = x^2$.

If you answered No, stop and review the
difference-of-two-squares formula again.
Make sure that one set of parentheses contains
a + sign and the other set of parentheses
contains a − sign.

If you answered Problem 2 incorrectly, go
back and rework the problem using these
suggestions.

Do you see how $16x^2 - 81$ can be factored using the formula
$a^2 - b^2$? Yes _____ No _____

Factoring a Perfect-Square Trinomial—Problem 4 Factor completely. $9a^2 - 30a + 25$

> **Helpful Hint:** Remember the perfect-square-trinomial formula: $a^2 - 2ab + b^2 = (a-b)^2$. You must verify
> two things to determine if you can use this formula:
> (1) The numerical values in the first term and the last term must be perfect squares.
> (2) The middle term must equal "twice the product of the values whose squares are the first and last terms."

Did you remember that $(3a)^2 = 9a^2$? Yes _____ No _____

Did you remember that $5^2 = 25$? Yes _____ No _____

If you answered No to these questions, stop and review the
first ten perfect squares. Consider that $3^2 = 9$ and $a \cdot a = a^2$.

If you answered No, check to see if the middle
term, $30a$, equals twice the product of $3a$
and 5.

Now go back and rework the problem using
these suggestions.

Do you see how $9a^2 - 30a + 25$ can be factored using the
formula $(a-b)^2$? Yes _____ No _____

Factoring a Polynomial with Four Terms by Grouping—Problem 6

Factor completely. $10xy + 15by - 8x - 12b$

> **Helpful Hint:** Look for common factors first. We can find the greatest common factors of the first two terms and factor. Then we can find the greatest common factor of the second two terms and factor. Make sure that you obtain the same binomial factor for each step. Be careful with $+/-$ signs.

Did you identify $5y$ as the greatest common factor of the first two terms: $10xy + 15by$?

Yes _____ No _____

If you answered No, remember that to factor completely, we must remove all common factors from both terms. Stop now and complete this step.

Did you identify 4 as the greatest common factor of the second two terms: $-8x - 12b$?

Yes _____ No _____

If you answered Problem 6 incorrectly, go back and rework the problem using these suggestions.

If you answered Yes to these questions, then you obtained $(2x + 3b)$ in the first term and $(-2x - 3b)$ in the second term. These are not the same binomial factor. Stop and consider how to get the same binomial factor of $(2x + 3b)$.

In your final answer, is the binomial factor of $(2x + 3b)$ listed once?

Yes _____ No _____

Factoring a Polynomial with a Common Factor—Problem 19

Factor completely. $3x^2 - 3x - 90$

> **Helpful Hint:** Look for the greatest common factor of all three terms as your first step. Don't forget to include this common factor as part of your answer. Always check your final product to make sure that it matches the original polynomial. Do this by multiplying.

Did you obtain $(3x - 18)(x + 5)$ or $(3x - 15)(x - 6)$ as your answer?

Yes _____ No _____

Now go back and rework the problem using these suggestions.

If you answered Yes, then you forgot to factor out the greatest common factor 3 as your first step.

Do you see how to factor $x^2 - x - 30$?

Yes _____ No _____

If you answered No, remember that we are looking for two numbers with a product of -30 and a sum of -1.

Be sure to double-check your final answer to be sure there are no common factors and include 3 in your final answer.

Chapter 6 Rational Expressions and Equations
6.1 Simplifying Rational Expressions

Vocabulary
rational expression   •   fractional algebraic expression   •   basic rule of fractions
simplify the fraction   •   factors

1. Only _____ of both the numerator and the denominator can be divided out.

2. A _____ is a polynomial divided by another polynomial.

3. Dividing out common factors is how to _____.

4. The _____ states that for any rational expression $\frac{a}{b}$ and any polynomials a, b,

    and c, $\frac{ac}{bc} = \frac{a}{b}$.

Example	Student Practice
1. Reduce. $\frac{21}{39}$	**2.** Reduce. $\frac{56}{77}$
Use the rule $\frac{ac}{bc} = \frac{a}{b}$. Let $c = 3$.	
$\frac{21}{39} = \frac{7 \cdot \cancel{3}}{13 \cdot \cancel{3}} = \frac{7}{13}$	
3. Simplify. $\frac{4x+12}{5x+15}$	**4.** Simplify. $\frac{6x+8}{15x+20}$
$\frac{4x+12}{5x+15} = \frac{4(x+3)}{5(x+3)}$	
$= \frac{4\cancel{(x+3)}}{5\cancel{(x+3)}}$	
$= \frac{4}{5}$	

Vocabulary Answers: 1. factors 2. rational expression 3. simplify the fraction 4. basic rule of fractions

Example	Student Practice
5. Simplify. $\dfrac{x^2+9x+14}{x^2-4}$	**6.** Simplify. $\dfrac{x^2+3x-28}{x^2-16}$

$$\frac{x^2+9x+14}{x^2-4}=\frac{(x+7)(x+2)}{(x-2)(x+2)}$$

$$=\frac{(x+7)\cancel{(x+2)}}{(x-2)\cancel{(x+2)}}$$

$$=\frac{x+7}{x-2}$$

7. Simplify. $\dfrac{x^3-9x}{x^3+x^2-6x}$	**8.** Simplify. $\dfrac{x^3+8x^2+12x}{x^3-36x}$

$$\frac{x^3-9x}{x^3+x^2-6x}=\frac{x(x^2-9x)}{x(x^2+x-6)}$$

$$=\frac{\cancel{x}\,\cancel{(x+3)}\,(x-3)}{\cancel{x}\,\cancel{(x+3)}\,(x-2)}$$

$$=\frac{x-3}{x-2}$$

9. Simplify. $\dfrac{5x-15}{6-2x}$	**10.** Simplify. $\dfrac{8x-24}{42-14x}$

The variable terms in the numerator and in the denominator are opposite in sign. Likewise the numerical terms are opposite in sign. Factor out a negative number from the denominator.

$$\frac{5x-15}{6-2x}=\frac{5(x-3)}{-2(-3+x)}$$

$$=\frac{5\cancel{(x-3)}}{-2\cancel{(-3+x)}}$$

$$=-\frac{5}{2}$$

Example	Student Practice
11. Simplify. $\dfrac{2x^2 - 11x + 12}{16 - x^2}$	**12.** Simplify. $\dfrac{6x^2 - 7x - 20}{25 - 4x^2}$

Factor the numerator and the denominator. Observe that $(x - 4)$ and $(4 - x)$ are opposites.

$$\frac{2x^2 - 11x + 12}{16 - x^2} = \frac{(x - 4)(2x - 3)}{(4 - x)(4 + x)}$$

Factor -1 out of $(+4 - x)$ to obtain $-1(-4 + x)$.

$$\frac{(x - 4)(2x - 3)}{(4 - x)(4 + x)} = \frac{(x - 4)(2x - 3)}{-1(-4 + x)(4 + x)}$$

$$= \frac{\cancel{(x - 4)}(2x - 3)}{-1\cancel{(-4 + x)}(4 + x)}$$

$$= \frac{(2x - 3)}{-1(4 + x)}$$

$$= -\frac{2x - 3}{4 + x}$$

Example	Student Practice
13. Simplify. $\dfrac{x^2 - 7xy + 12y^2}{2x^2 - 7xy - 4y^2}$	**14.** Simplify. $\dfrac{10x^2 - 9xy - 9y^2}{18x^2 - 13xy - 21y^2}$

$$\frac{x^2 - 7xy + 12y^2}{2x^2 - 7xy - 4y^2} = \frac{(x - 4y)(x - 3y)}{(2x + y)(x - 4y)}$$

$$= \frac{\cancel{(x - 4y)}(x - 3y)}{(2x + y)\cancel{(x - 4y)}}$$

$$= \frac{x - 3y}{2x + y}$$

Example	Student Practice
15. Simplify. $\dfrac{6a^2 + ab - 7b^2}{36a^2 - 49b^2}$	**16.** Simplify. $\dfrac{25a^2 - 16b^2}{10a^2 - 3ab - 4b^2}$

$$\frac{6a^2 + ab - 7b^2}{36a^2 - 49b^2} = \frac{(6a + 7b)(a - b)}{(6a + 7b)(6a - 7b)}$$

$$= \frac{\cancel{(6a + 7b)}(a - b)}{\cancel{(6a + 7b)}(6a - 7b)}$$

$$= \frac{a - b}{6a - 7b}$$

Extra Practice

1. Simplify. $\dfrac{3x + 12}{x^2 + 4x}$

2. Simplify. $\dfrac{9x^2 - 24x + 16}{9x^2 - 16}$

3. Simplify. $\dfrac{81 - x^2}{3x^2 - 21x - 54}$

4. Simplify. $\dfrac{25x^2 - 20xy + 4y^2}{15x^2 - xy - 2y^2}$

Concept Check

Explain why it is important to completely factor both the numerator and the denominator when simplifying $\dfrac{x^2 y - y^3}{x^2 y + xy^2 - 2y^3}$.

Chapter 6 Rational Expressions and Equations
6.2 Multiplying and Dividing Rational Expressions

Vocabulary

reciprocals   •   multiply   •   divide   •   greatest common factor

1. You should always check for the _____ as your first step when multiplying rational expressions.

2. Two numbers are _____ of each other if their product is 1.

3. To _____ two rational expressions, multiply the numerators and multiply the denominators.

4. To _____ two rational expressions, invert the second fraction and multiply it by the first fraction.

Example	Student Practice
1. Multiply. $\dfrac{x^2-x-12}{x^2-16}\cdot\dfrac{2x^2+7x-4}{x^2-4x-21}$	**2.** Multiply. $\dfrac{3x^2+34x+63}{2x^2+7x-15}\cdot\dfrac{x^2+9x+20}{x^2+13x+36}$

Factoring is always the first step.

$$\frac{x^2-x-12}{x^2-16}\cdot\frac{2x^2+7x-4}{x^2-4x-21}$$
$$=\frac{(x-4)(x+3)}{(x-4)(x+4)}\cdot\frac{(x+4)(2x-1)}{(x+3)(x-7)}$$

Apply the basic rule of fractions. Three pairs of factors divide out.

$$\frac{(x-4)(x+3)}{(x-4)(x+4)}\cdot\frac{(x+4)(2x-1)}{(x+3)(x-7)}$$
$$=\frac{\cancel{(x-4)}\,\cancel{(x+3)}}{\cancel{(x-4)}\,(x+4)}\cdot\frac{\cancel{(x+4)}\,(2x-1)}{\cancel{(x+3)}\,(x-7)}$$
$$=\frac{(2x-1)}{(x-7)}$$

Vocabulary Answers: 1. greatest common factor 2. reciprocals 3. multiply 4. divide

Example	Student Practice
3. Multiply. $\dfrac{x^4-16}{x^3+4x}\cdot\dfrac{2x^2-8x}{4x^2+2x-12}$	**4.** Multiply. $\dfrac{x^3+x}{2x^4-2}\cdot\dfrac{4x^3-4x}{6x^3+38x^2+40x}$

Factor each numerator and denominator. Factoring out the greatest common factor first is very important.

$$\frac{x^4-16}{x^3+4x}\cdot\frac{2x^2-8x}{4x^2+2x-12}$$

$$=\frac{\left(x^2+4\right)\left(x^2-4\right)}{x\left(x^2+4\right)}\cdot\frac{2x(x-4)}{2\left(2x^2+x-6\right)}$$

$$=\frac{\left(x^2+4\right)(x+2)(x-2)}{x\left(x^2+4\right)}\cdot\frac{2x(x-4)}{2(x+2)(2x-3)}$$

$$=\frac{\cancel{\left(x^2+4\right)}\cancel{(x+2)}(x-2)}{\cancel{x}\cancel{\left(x^2+4\right)}}\cdot\frac{\cancel{2}\cancel{x}(x-4)}{\cancel{2}\cancel{(x+2)}(2x-3)}$$

$$=\frac{(x-2)(x-4)}{2x-3}\ \text{or}\ \frac{x^2-6x+8}{2x-3}$$

---

5. Divide. $\dfrac{6x+12y}{2x-6y}\div\dfrac{9x^2-36y^2}{4x^2-36y^2}$

6. Divide.
$$\frac{5x^2+20x-105}{4x^2-44x+96}\div\frac{10x^2+75x+35}{16x^2-140x+96}$$

$$=\frac{6x+12y}{2x-6y}\cdot\frac{4x^2-36y^2}{9x^2-36y^2}$$

$$=\frac{6(x+2y)}{2(x-3y)}\cdot\frac{4\left(x^2-9y^2\right)}{9\left(x^2-4y^2\right)}$$

$$=\frac{(3)(2)(x+2y)}{2(x-3y)}\cdot\frac{(2)(2)(x+3y)(x-3y)}{(3)(3)(x+2y)(x-2y)}$$

$$=\frac{\cancel{(3)}\cancel{(2)}\cancel{(x+2y)}}{\cancel{2}\cancel{(x-3y)}}\cdot\frac{(2)(2)(x+3y)\cancel{(x-3y)}}{\cancel{(3)}(3)\cancel{(x+2y)}(x-2y)}$$

$$=\frac{(2)(2)(x+3y)}{3(x-2y)}$$

$$=\frac{4(x+3y)}{3(x-2y)}$$

178

Example	Student Practice
7. Divide. $\dfrac{15-3x}{x+6} \div \left(x^2 - 9x + 20\right)$	**8.** Divide. $\dfrac{x+4}{x-2} \div \left(-x^2 + x + 20\right)$

Note that $x^2 - 9x + 20$ can be written as $\dfrac{x^2 - 9x + 20}{1}$.

$$\dfrac{15-3x}{x+6} \div \left(x^2 - 9x + 20\right)$$

$$= \dfrac{15-3x}{x+6} \cdot \dfrac{1}{x^2 - 9x + 20}$$

$$= \dfrac{-3(-5+x)}{x+6} \cdot \dfrac{1}{(x-5)(x-4)}$$

$$= \dfrac{-3(\cancel{-5+x})}{x+6} \cdot \dfrac{1}{\cancel{(x-5)}(x-4)}$$

$$= \dfrac{-3}{(x+6)(x-4)}$$

or $-\dfrac{3}{(x+6)(x-4)}$ or $\dfrac{3}{(x+6)(4-x)}$

Extra Practice

1. Multiply. $\dfrac{32x^3}{8x^2 - 8} \cdot \dfrac{4x-4}{16x^2}$

2. Multiply.
$$\dfrac{2x^2 - 3x}{4x^3 + 4x^2 - 9x - 9} \cdot \dfrac{2x^2 + 5x + 3}{5x+7}$$

3. Divide. $\dfrac{9x^2 - 49}{3x^2 + 8x - 35} \div \left(3x^2 + x - 14\right)$

4. Divide. $\dfrac{9x^2 - 16}{9x^2 + 24x + 16} \div \dfrac{3x^2 - x - 4}{5x^2 - 5}$

Concept Check

Explain how you would divide $\dfrac{21x-7}{9x^2-1} \div \dfrac{1}{3x+1}$.

Name: _____   Date: _____

Instructor: _____   Section: _____

Chapter 6 Rational Expressions and Equations
6.3 Adding and Subtracting Rational Expressions

Vocabulary
least common denominator   •   denominator   •   different   •   factor

1. If rational expressions have the same _____, they can be combined in the same way as arithmetic fractions.

2. If two rational expressions have _____ denominators, we first change them to equivalent rational expressions with the least common denominator.

3. The _____ is a product containing each different factor for each denominator of rational expressions.

4. The first step to find the LCD of two or more rational expressions is to _____ each denominator completely.

Example	Student Practice
1. Add. $\dfrac{5a}{a+2b}+\dfrac{6a}{a+2b}$	**2.** Add. $\dfrac{4x+3}{3x+4}+\dfrac{5-4x}{3x+4}$
Note that the denominators are the same. Only add the numerators. Do not change the denominator.	
$\dfrac{5a}{a+2b}+\dfrac{6a}{a+2b}=\dfrac{5a+6a}{a+2b}=\dfrac{11a}{a+2b}$	
3. Subtract. $\dfrac{3x}{(x+y)(x-2y)}-\dfrac{8x}{(x+y)(x-2y)}$	**4.** Subtract. $\dfrac{4x+6}{(3x-2)(x+9)}-\dfrac{2x-5}{(3x-2)(x+9)}$
Write as one fraction and simplify.	
$\dfrac{3x}{(x+y)(x-2y)}-\dfrac{8x}{(x+y)(x-2y)}$	
$=\dfrac{3x-8x}{(x+y)(x-2y)}=\dfrac{-5x}{(x+y)(x-2y)}$	

Vocabulary Answers: 1. denominator 2. different 3. least common denominator 4. factor

Example	Student Practice
5. Find the LCD. $\dfrac{5}{2x-4}, \dfrac{6}{3x-6}$	**6.** Find the LCD. $\dfrac{5}{16x+20}, \dfrac{11}{28x+35}$

Factor each denominator.

$$2x-4 = 2(x-2)$$
$$3x-6 = 3(x-2)$$

The factors are 2, 3, and $(x-2)$. The LCD is the product of these factors.

$$LCD = (2)(3)(x-2) = 6(x-2)$$

---

7. Find the LCD.

$$\frac{5}{12ab^2c}, \frac{13}{18a^3bc^4}$$

The LCD will contain each factor repeated the greatest number of times that it occurs in any one denominator.

$12ab^2x = 2\cdot2\cdot3 \quad\cdot a \qquad \cdot b\cdot b\cdot c$

$18a^3bc^4 = \quad 2\cdot3\cdot3\cdot a\cdot a\cdot a\cdot b\cdot \quad c\cdot c\cdot c\cdot c$

$LCD = 2\cdot2\cdot3\cdot3\cdot a\cdot a\cdot a\cdot b\cdot b\cdot c\cdot c\cdot c\cdot c$

$LCD = 2^2\cdot3^2\cdot a^3\cdot b^2\cdot c^4 = 36a^3b^2c^4$

8. Find the LCD.

(a) $\dfrac{7}{24ab^2c^3}, \dfrac{13}{36a^2b^4c}$

(b) $\dfrac{11}{2x^2+17x+30}, \dfrac{9}{x^2+15x+54}$

---

9. Add. $\dfrac{5}{xy} + \dfrac{2}{y}$

10. Add. $\dfrac{4}{xyz} + \dfrac{2}{y}$

Find the LCD. The two factors are x and y. Observe that the LCD is xy.

$$\frac{5}{xy} + \frac{2}{y} = \frac{5}{xy} + \frac{2}{y}\cdot\frac{x}{x}$$

$$= \frac{5}{xy} + \frac{2x}{xy}$$

$$= \frac{5+2x}{xy}$$

Example	Student Practice
11. Add. $\dfrac{3x}{x^2-y^2}+\dfrac{5}{x+y}$	**12.** Add. $\dfrac{4}{2x-5}+\dfrac{3x+2}{4x^2-25}$

Factor the first denominator so that $x^2-y^2=(x+y)(x-y)$. Thus, the factors of the denominators are $(x+y)$ and $(x-y)$. Observe that the LCD $=(x+y)(x-y)$.

$\dfrac{3x}{x^2-y^2}+\dfrac{5}{x+y}$

$=\dfrac{3x}{(x+y)(x-y)}+\dfrac{5}{(x+y)}\cdot\dfrac{x-y}{x-y}$

$=\dfrac{3x}{(x+y)(x-y)}+\dfrac{5x-5y}{(x+y)(x-y)}$

$=\dfrac{8x-5y}{(x+y)(x-y)}$

Example	Student Practice
13. Add. $\dfrac{5}{x^2-y^2}+\dfrac{3x}{x^3+x^2y}$	**14.** Add. $\dfrac{6y}{3xy-y^2}+\dfrac{4y}{9x^2-y^2}$

$\dfrac{5}{x^2-y^2}+\dfrac{3x}{x^3+x^2y}$

$=\dfrac{5}{(x+y)(x-y)}+\dfrac{3x}{x^2(x+y)}$

$=\dfrac{5}{(x+y)(x-y)}\cdot\dfrac{x^2}{x^2}+\dfrac{3x}{x^2(x+y)}\cdot\dfrac{x-y}{x-y}$

$=\dfrac{5x^2}{x^2(x+y)(x-y)}+\dfrac{3x^2-3xy}{x^2(x+y)(x-y)}$

$=\dfrac{x(8x-3y)}{x^2(x+y)(x-y)}$

$=\dfrac{8x-3y}{x(x+y)(x-y)}$

Example	Student Practice

Example

15. Subtract. $\dfrac{3x+4}{x-2} - \dfrac{x-3}{2x-4}$

Factor the second denominator.

$$\frac{3x+4}{x-2} - \frac{x-3}{2x-4} = \frac{3x+4}{x-2} - \frac{x-3}{2(x-2)}$$

$$= \frac{2}{2} \cdot \frac{3x+4}{x-2} - \frac{x-3}{2(x-2)}$$

$$= \frac{2(3x+4)-(x-3)}{2(x-2)}$$

$$= \frac{6x+8-x+3}{2(x-2)}$$

$$= \frac{5x+11}{2(x-2)}$$

Student Practice

16. Subtract. $\dfrac{x+5}{2x+7} - \dfrac{x-2}{6x+21}$

Extra Practice

1. Find the LCD. Do not combine fractions.
$$\frac{7}{2x^2-11x+12}, \ \frac{13}{2x^2+7x-15}$$

2. Perform the operation indicated. Be sure to simplify. $\dfrac{5x}{x+1} - \dfrac{2x-5}{x+1}$

3. Perform the operation indicated. Be sure to simplify. $\dfrac{7}{x^2-y^2} + \dfrac{2y}{xy^2+y^3}$

4. Perform the operation indicated. Be sure to simplify. $\dfrac{1}{x^2+3x+2} - \dfrac{2}{x^2-x-6}$

Concept Check

Explain how to find the LCD of the fractions $\dfrac{3}{10xy^2z}$ and $\dfrac{6}{25x^2yz^3}$.

Name: _____   Date: _____

Instructor: _____   Section: _____

Chapter 6 Rational Expressions and Equations
6.4 Simplifying Complex Rational Expressions

Vocabulary

complex rational expression • complex fraction • numerator • denominator • LCD

1. A complex rational expression is also called a(n) _____.

2. A(n) _____ has a fraction in the numerator or in the denominator, or both.

3. A complex rational expression may contain two or more fractions in the _____ and denominator.

4. To simplify complex rational expressions, multiply the numerator and denominator of the complex fraction by the _____ of all the denominators appearing in the complex fraction.

Example	Student Practice
1. Simplify. $\dfrac{\dfrac{1}{x}}{\dfrac{2}{y^2}+\dfrac{1}{y}}$	**2.** Simplify. $\dfrac{\dfrac{1}{n^2}}{\dfrac{1}{m}+\dfrac{3}{4}}$

Add the two fractions in the denominator.

$$\frac{\dfrac{1}{x}}{\dfrac{2}{y^2}+\dfrac{1}{y}\cdot\dfrac{y}{y}}=\frac{\dfrac{1}{x}}{\dfrac{2+y}{y^2}}$$

Divide the fraction in the numerator by the fraction in the denominator.

$$\frac{1}{x}\div\frac{2+y}{y^2}=\frac{1}{x}\cdot\frac{y^2}{2+y}$$

$$=\frac{y^2}{x(2+y)}$$

Vocabulary Answers: 1. complex fraction 2. complex rational expression 3. numerator 4. LCD

Example	**Student Practice**

3. Simplify. $\dfrac{\dfrac{1}{x}+\dfrac{1}{y}}{\dfrac{3}{a}-\dfrac{2}{b}}$

4. Simplify. $\dfrac{\dfrac{1}{m}+\dfrac{1}{n}}{\dfrac{x}{5}-\dfrac{y}{3}}$

Observe that the LCD of the fractions in the numerator is xy. The LCD of the fractions in the denominator is ab.

$$\dfrac{\dfrac{1}{x}+\dfrac{1}{y}}{\dfrac{3}{a}-\dfrac{2}{b}}=\dfrac{\dfrac{1}{x}\cdot\dfrac{y}{y}+\dfrac{1}{y}\cdot\dfrac{x}{x}}{\dfrac{3}{a}\cdot\dfrac{b}{b}-\dfrac{2}{b}\cdot\dfrac{a}{a}}=\dfrac{\dfrac{y+x}{xy}}{\dfrac{3b-2a}{ab}}$$

$$=\dfrac{y+x}{xy}\cdot\dfrac{ab}{3b-2a}=\dfrac{ab(y+x)}{xy(3b-2a)}$$

5. Simplify. $\dfrac{\dfrac{1}{x^2-1}+\dfrac{2}{x+1}}{x}$

6. Simplify. $\dfrac{\dfrac{3x}{x^2+8x+12}-\dfrac{2}{x+6}}{5}$

We need to factor x^2-1.

$$\dfrac{\dfrac{1}{x^2-1}+\dfrac{2}{x+1}}{x}$$

$$=\dfrac{\dfrac{1}{(x+1)(x-1)}+\dfrac{2}{x+1}\cdot\dfrac{x-1}{x-1}}{x}$$

$$=\dfrac{\dfrac{1+2x-2}{(x+1)(x-1)}}{x}$$

$$=\dfrac{2x-1}{(x+1)(x-1)}\cdot\dfrac{1}{x}$$

$$=\dfrac{2x-1}{x(x+1)(x-1)}$$

Example	Student Practice

Example

7. Simplify. $\dfrac{\dfrac{3}{a+b} - \dfrac{3}{a-b}}{\dfrac{5}{a^2-b^2}}$

$$= \dfrac{\dfrac{3}{a+b}\cdot\dfrac{a-b}{a-b} - \dfrac{3}{a-b}\cdot\dfrac{a+b}{a+b}}{\dfrac{5}{a^2-b^2}}$$

$$= \dfrac{\dfrac{3a-3b}{(a+b)(a-b)} - \dfrac{3a+3b}{(a+b)(a-b)}}{\dfrac{5}{a^2-b^2}}$$

$$= \dfrac{\dfrac{-6b}{(a+b)(a-b)}}{\dfrac{5}{(a+b)(a-b)}}$$

$$= \dfrac{-6b}{(a+b)(a-b)}\cdot\dfrac{(a+b)(a-b)}{5}$$

$$= \dfrac{-6b}{5} \text{ or } -\dfrac{6b}{5}$$

Student Practice

8. Simplify. $\dfrac{\dfrac{5}{x+y} + \dfrac{2}{x-y}}{\dfrac{7}{x^2-y^2}}$

---

9. Simplify by multiplying by the LCD.

$$\dfrac{\dfrac{5}{ab^2} - \dfrac{2}{ab}}{3 - \dfrac{5}{2a^2b}}$$

$$= \dfrac{2a^2b^2\left(\dfrac{5}{ab^2} - \dfrac{2}{ab}\right)}{2a^2b^2\left(3 - \dfrac{5}{2a^2b}\right)}$$

$$= \dfrac{2a^2b^2\left(\dfrac{5}{ab^2}\right) - 2a^2b^2\left(\dfrac{2}{ab}\right)}{2a^2b^2(3) - 2a^2b^2\left(\dfrac{5}{2a^2b}\right)}$$

$$= \dfrac{10a - 4ab}{6a^2b^2 - 5b}$$

10. Simplify by multiplying by the LCD.

$$\dfrac{\dfrac{4}{y} - \dfrac{7}{4x}}{5 - \dfrac{9}{xy^2}}$$

Example	Student Practice

Example

11. Simplify by multiplying by the LCD.

$$\frac{\dfrac{3}{a+b}-\dfrac{3}{a-b}}{\dfrac{5}{a^2-b^2}}$$

The LCD of all individual fractions in the complex fraction is $(a+b)(a-b)$.

$$=\frac{(a+b)(a-b)\left(\dfrac{3}{a+b}\right)-(a+b)(a-b)\left(\dfrac{3}{a-b}\right)}{(a+b)(a-b)\left(\dfrac{5}{(a+b)(a-b)}\right)}$$

$$=\frac{3(a-b)-3(a+b)}{5}$$

$$=\frac{3a-3b-3a-3b}{5}$$

$$=-\frac{6b}{5}$$

Student Practice

12. Simplify by multiplying by the LCD.

$$\frac{\dfrac{5}{x+y}+\dfrac{2}{x-y}}{\dfrac{7}{x^2-y^2}}$$

Extra Practice

1. Simplify. $\dfrac{\dfrac{4}{a}+\dfrac{5}{b}}{\dfrac{2}{ab}}$

2. Simplify. $\dfrac{\dfrac{2}{x}+\dfrac{2}{y}}{x+y}$

3. Simplify. $\dfrac{\dfrac{2x}{x^2-25}}{\dfrac{4}{x+5}-\dfrac{3}{x-5}}$

4. Simplify. $\dfrac{\dfrac{3}{4x}-\dfrac{8}{3y}}{\dfrac{7}{a}+\dfrac{5}{b}}$

Concept Check

To simplify the following complex fraction, explain how you would add the two fractions in the numerator.

$$\frac{\dfrac{7}{x-3}+\dfrac{15}{2x-6}}{\dfrac{2}{x+5}}$$

Chapter 6 Rational Expressions and Equations
6.5 Solving Equations Involving Rational Expressions

Vocabulary
extraneous solution • no solution • LCD • exclude

1. The first step to solve an equation containing rational expressions is to determine the _____ of all the denominators.

2. If all of the apparent solutions of an equation are extraneous solutions, we say that the equation has _____.

3. A value that makes a denominator in the equation equal to zero is called a(n) _____.

4. _____ from your solution any value that would make the LCD equal to zero.

Example	Student Practice
1. Solve for x and check your solution. $$\frac{5}{x}+\frac{2}{3}=-\frac{3}{x}$$	**2.** Solve for x and check your solution. $$\frac{5}{x}+\frac{1}{4}=-\frac{6}{x}$$

The LCD is $3x$. Multiply each term by $3x$.

$$3x\left(\frac{5}{x}\right)+3x\left(\frac{2}{3}\right)=3x\left(-\frac{3}{x}\right)$$

$$15+2x=-9$$
$$2x=-9-15$$
$$2x=-24$$
$$x=-12$$

Check. Replace each x by -12.

$$\frac{5}{-12}+\frac{2}{3}\overset{?}{=}-\frac{3}{-12}$$

$$-\frac{5}{12}+\frac{8}{23}\overset{?}{=}\frac{3}{12}$$

$$\frac{3}{12}=\frac{3}{12}$$

Vocabulary Answers: 1. LCD 2. no solution 3. extraneous solution 4. exclude

Example	Student Practice
3. Solve and check. $\dfrac{6}{x+3} = \dfrac{3}{x}$	**4.** Solve and check. $\dfrac{2}{x+4} = \dfrac{6}{x}$

Observe that the LCD $= x(x+3)$.

$$x(x+3)\left(\frac{6}{x+3}\right) = x(x+3)\left(\frac{3}{x}\right)$$
$$6x = 3(x+3)$$
$$6x = 3x+9$$
$$3x = 9$$
$$x = 3$$

Check. Replace each x by 3.

$$\frac{6}{3+3} \overset{?}{=} \frac{3}{3}$$
$$\frac{6}{6} = \frac{3}{3}$$

---

5. Solve and check. $\dfrac{3}{x+5} - 1 = \dfrac{4-x}{2x+10}$

6. Solve and check.

$$\frac{3}{x+2} - \frac{5}{x+5} = \frac{x-2}{x^2+7x+10}$$

$$\frac{3}{x+5} - 1 = \frac{4-x}{2(x+5)}$$

$$2(x+5)\left(\frac{3}{x+5}\right) - 2(x+5)(1)$$

$$= 2(x+5)\left[\frac{4-x}{2(x+5)}\right]$$

$$2(3) - 2(x+5) = 4-x$$
$$6 - 2x - 10 = 4-x$$
$$-2x - 4 = 4-x$$
$$-4 = 4+x$$
$$-8 = x$$

The check is left to the student.

Example	Student Practice
7. Solve and check. $\dfrac{y}{y-2} - 4 = \dfrac{2}{y-2}$	**8.** Solve and check. $\dfrac{3x}{x-3} + 5 = \dfrac{3x}{x-3}$

Observe that the LCD is $y - 2$.

$$(y-2)\left(\frac{y}{y-2}\right) - (y-2)(4)$$

$$= (y-2)\left(\frac{2}{y-2}\right)$$

$$y - 4(y-2) = 2$$

$$y - 4y + 8 = 2$$

$$-3y + 8 = 2$$

$$-3y = -6$$

$$\frac{-3y}{-3} = \frac{-6}{-3}$$

$$y = 2$$

This equation has no solution. We can see immediately that $y = 2$ is not a solution of the original equation. When we substitute 2 for y in a denominator, the denominator is equal to zero and the expression is undefined. Suppose that we tried to check the apparent solution by substituting 2 for y.

$$\frac{y}{y-2} - 4 = \frac{2}{y-2}$$

$$\frac{2}{2-2} - 4 \overset{?}{=} \frac{2}{2-2}$$

$$\frac{2}{0} - 4 = \frac{2}{0}$$

These expressions are not defined. There is no such number as $2 \div 0$. We see that 2 does not check. This equation has no solution.

Extra Practice

1. Solve and check. If there is no solution, so indicate. $\dfrac{2}{3x} + \dfrac{1}{6} = \dfrac{6}{x}$

2. Solve and check. If there is no solution, so indicate. $\dfrac{4}{a^2 - 1} = \dfrac{2}{a+1} + \dfrac{2}{a-1}$

3. Solve and check. If there is no solution, so indicate.
$$\dfrac{6}{5a + 10} - \dfrac{1}{a - 5} = \dfrac{4}{a^2 - 3a - 10}$$

4. Solve and check. If there is no solution, so indicate.
$$\dfrac{x-2}{x^2 - 4x - 5} + \dfrac{x+5}{x^2 - 25} = \dfrac{2x+13}{x^2 + 6x + 5}$$

Concept Check

Explain how to find the LCD for the following equation. Do not solve the equation.
$$\dfrac{x}{x^2 - 9} + \dfrac{2}{3x - 9} = \dfrac{5}{2x + 6} + \dfrac{3}{2x^2 - 18}$$

Chapter 6 Rational Expressions and Equations
6.6 Ratio, Proportion, and Other Applied Problems

Vocabulary
ratio   •   proportion   •   cross multiplying   •   similar

1. A _____ is an equation that states that two ratios are equal.

2. _____ triangles are triangles that have the same shape, but may be different sizes.

3. If $\dfrac{a}{b} = \dfrac{c}{d}$, then $ad = bc$ is sometimes called _____.

4. A _____ is a comparison of two quantities.

Example	Student Practice
1. Michael took 5 hours to drive 245 miles on the turnpike. At the same rate, how many hours will it take him to drive a distance of 392 miles?	**2.** Rose took 4 hours to drive 264 miles. How far will she drive in 9 hours?

Let x = the number of hours it will take to drive 392 miles. If 5 hours are needed to drive 245 miles, then x hours are needed to drive 392 miles.

$$\frac{5 \text{ hours}}{245 \text{ miles}} = \frac{x \text{ hours}}{392 \text{ miles}}$$

$$5(392) = 245x$$

$$\frac{1960}{245} = x$$

$$8 = x$$

It will take Michael 8 hours to drive 392 miles. To check, do the computations and see if $\dfrac{5}{245} = \dfrac{8}{392}$.

Vocabulary Answers: 1. proportion 2. similar 3. cross multiplying 4. ratio

Example	Student Practice
3. A ramp is 32 meters long and rises up 15 meters. A ramp at the same angle is 9 meters long. How high does the second ramp rise?	**4.** Triangle M is similar to Triangle N. Find the length of side x. Express your answer as a mixed number.

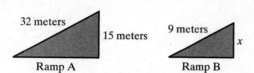

Ramp A   Ramp B

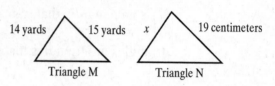

Triangle M   Triangle N

$$\frac{32}{9} = \frac{15}{x}$$

$$32x = (9)(15)$$

$$32x = 135$$

$$x = \frac{135}{32} \text{ or } x = 4\frac{7}{32}$$

---

5. Plane A flies at a speed that is 50 kilometers per hour faster than plane B. Plane A flies 500 kilometers in the amount of time that plane B flies 400 kilometers. Find the speed of each plane.

Let $s =$ the speed of plane B in kilometers per hour. Then $s + 50 =$ the speed of plane A in kilometers per hour. Each plane flies the same amount of time. That is, the time for plane A equals the time for plane B.

$$\frac{500}{s+50} = \frac{400}{s}$$

$$500s = (s+50)(400)$$

$$500s = 400s + 20,000$$

$$100s = 20,000$$

$$s = 200$$

Plane B travels 200 kilometers per hour. Since $s + 50 = 200 + 50 = 250$, plane A travels 250 kilometers per hour.

6. Two trains travel in opposite directions for the same amount of time. Train A traveled 200 kilometers, while train B traveled 175 kilometers. Train A traveled 15 kilometers per hour faster than train B. What is the speed of each train?

Example	Student Practice

Example

7. Reynaldo can sort a huge stack of mail in 9 hours. His brother Carlos can sort the mail in 8 hours. How long will it take them to do the job together? Express your answer in hours and minutes. Round to the nearest minute.

If Reynaldo can do the job in 9 hours, then in 1 hour he could do $\frac{1}{9}$ of the job. If Carlos can do the job in 8 hours, then in 1 hour he could do $\frac{1}{8}$ of the job. Let x = the number of hours it takes Reynaldo and Carlos to do the job together. In 1 hour together they could do $\frac{1}{x}$ of the job. The amount of work Reynaldo can do in 1 hour plus the amount of work Carlos can do in 1 hour must be equal to the amount of work they could do together in 1 hour.

$$\frac{1}{9}+\frac{1}{8}=\frac{1}{x}$$

$$72x\left(\frac{1}{9}\right)+72x\left(\frac{1}{8}\right)=72x\left(\frac{1}{x}\right)$$

$$8x+9x=72$$

$$17x=72$$

$$x=4\frac{4}{17}$$

To change $\frac{4}{17}$ of an hour to minutes,

$$\frac{4}{17}\ \text{hour}\times\frac{60\ \text{minutes}}{1\ \text{hour}}=\frac{240}{17}\ \text{minutes},$$

which is approximately 14.118. Thus doing the job together will take 4 hours and 14 minutes.

Student Practice

8. Two people are painting a side of a house. The first person takes 4 hours to paint the side of the house, while the second person takes 6 hours. How long will it take both of them to paint the side of the house working together? Express your answer in hours and minutes. Round to the nearest minute.

195

Extra Practice

1. Solve. $\dfrac{3}{10} = \dfrac{11.4}{x}$

2. The scale on a map of Massachusetts is approximately $\dfrac{3}{4}$ inch to 25 miles. If the distance from a college to Boston measures 4 inches on the map, how far apart are the two locations?

3. Alicia is 4 feet tall and casts a shadow that is 9 feet long. At the same time of day, a tree casts a shadow that is 36 feet long. How tall is the tree?

4. Toshiro and Mathilda work in the library. Working alone, it takes Toshiro 7 hours to arrange and stack 1000 books, while it takes Mathilda 6.5 hours. To the nearest minute, how long will the job take if they work together to arrange and stack 1000 books?

Concept Check

Mike found that his car used 18 gallons of gas to travel 396 miles. He needs to take a trip of 450 miles and wants to know how many gallons of gas it will take. He set up the equation $\dfrac{18}{x} = \dfrac{450}{396}$. Explain what error he made and how he should correctly solve the problem.

MATH COACH

Mastering the skills you need to do well on the test.

Watch the **MATH COACH** videos in MyMathLab® or on You[Tube]™ while you work the problems below. These helpful hints will help you avoid making common errors on test problems.

Dividing Rational Expressions—Problem 5 $\dfrac{2a^2-3a-2}{a^2+5a+6}\div\dfrac{a^2-5a+6}{a^2-9}$

> **Helpful Hint:** The operation of division is performed by inverting the second fraction and multiplying it by the first fraction. This step should be done first, before you begin factoring.

Did you keep the first fraction the same as it is written and then multiply it by $\dfrac{a^2-9}{a^2-5a+6}$? Yes _____ No _____
If you answered No, stop and make this correction to your work.

Were you able to factor each expression so that you obtained $\dfrac{(2a+1)(a-2)}{(a+2)(a+3)}\cdot\dfrac{(a+3)(a-3)}{(a-2)(a-3)}$? Yes _____ No _____

If you answered No, go back and check each factoring step to see if you can get the same result. Complete the problem by dividing out any common factors.

If you answered Problem 5 incorrectly, go back and rework the problem using these suggestions.

Subtracting Rational Expressions with Different Denominators—Problem 8 $\dfrac{3x}{x^2-3x-18}-\dfrac{x-4}{x-6}$

> **Helpful Hint:** First factor the trinomial in the denominator so that you can determine the LCD of these two fractions. Then multiply by what is needed in the second fraction so that it becomes an equivalent fraction with the LCD as the denominator. When subtracting, it is a good idea to place brackets around the second numerator to avoid sign errors.

Did you factor the first denominator into $(x-6)(x+3)$?
Yes _____ No _____

Did you then determine that the LCD is $(x-6)(x+3)$?
Yes _____ No _____

If you answered No to these questions, stop and review how to factor the trinomial in the first denominator and how to find the LCD when working with polynomials as denominators.

Did you multiply the numerator and denominator of the second fraction by $(x+3)$ and place brackets around the product to obtain $\dfrac{3x}{(x-6)(x+3)}-\dfrac{\left[(x-4)(x+3)\right]}{(x-6)(x+3)}$?

Yes _____ No _____

Did you obtain $3x-x^2-x-12$ in the numerator? Yes _____ No _____

If you answered Yes, you forgot to distribute the negative sign. Rework the problem and make this correction.

As your final step, remember to combine like terms and then factor the new numerator before dividing out common factors.

Now go back and rework the problem using these suggestions.

Simplify. $\dfrac{\dfrac{6}{b}-4}{\dfrac{5}{bx}-\dfrac{10}{3x}}$

Helpful Hint: There are two ways to simplify this expression:
1) combine the numerators and the denominators separately, or
2) multiply the numerator and denominator of the complex fraction by the LCD of all the denominators.
Consider both methods and choose the one that seems easiest to you. We will show the steps of the first method for this particular problem.

Did you multiply 4 by $\dfrac{b}{b}$ and then subtract $\dfrac{6}{b}-\dfrac{4b}{b}$?

Yes _____ No _____

If you answered No, remember that 4 can be written as $\dfrac{4}{1}$, and to find the LCD, you must multiply numerator and denominator by the variable b.

In the denominator of the complex fraction, did you obtain $3bx$ as the LCD and multiply the first fraction by $\dfrac{3}{3}$ and the second fraction by $\dfrac{b}{b}$ before subtracting the two fractions?

Yes _____ No _____

If you answered No, stop and carefully change these two fractions into equivalent fractions with $3bx$ as the

denominator. Then subtract the numerators and keep the common denominator.

Were you able to rewrite the problem as follows: $\dfrac{6-4b}{b}\div\dfrac{15-10b}{3bx}$?

Yes _____ No _____

If you answered No, examine your steps carefully. Once you have one fraction in the numerator and one fraction in the denominator, you can rewrite the division as multiplication by inverting the fraction in the denominator.

If you answered Problem 10 incorrectly, go back and rework the problem using these suggestions.

Solving Equations Involving Rational Expressions—Problem 14

Solve for x. $\dfrac{x-3}{x-2}=\dfrac{2x^2-15}{x^2+x-6}-\dfrac{x+1}{x+3}$

Helpful Hint: First factor any denominators that need to be factored so that you can determine the LCD of all denominators. Verify that you are solving an equation, then multiply each term of the equation by the LCD. Solve the resulting equation. Check your solution.

Did you factor x^2+x-6 to get $(x+3)(x-2)$ and identify the LCD as $(x+3)(x-2)$? Yes _____ No _____
If you answered No, go back and complete these steps again.

Did you notice that we are solving an equation and that we can multiply each term of the equation by the LCD and then multiply the binomials to obtain the equation $x^2-9=(2x^2-15)-(x^2-x-2)$? Yes _____ No _____

If you answered No, look at the step $-\left[(x+1)(x-2)\right]$.

After you multiplied the binomials, did you distribute the negative sign and change the sign of all terms inside the grouping symbols? Yes _____ No _____

If you answered No, review how to multiply binomials and subtract polynomial expressions. Then combine like terms and solve the equation for x.

Now go back and rework the problem using these suggestions.

Name: _____   Date: _____
Instructor: _____   Section: _____

Chapter 7 Graphing and Functions
7.1 The Rectangular Coordinate System

Vocabulary
graphs   •   rectangular coordinate system   •   origin   •   *x*-axis   •   *y*-axis
ordered pair   •   coordinates   •   *x*-coordinate   •   *y*-coordinate   •   solution

1. The numbers in an ordered pair are often referred to as the _____ of the point.

2. We can illustrate algebraic relationships with drawings called _____.

3. The vertical number line above the origin is often called the _____.

4. The first number in an ordered pair is called the _____ and it represents the distance from the origin measured along the horizontal axis.

Example	Student Practice
1. Plot the point $(5,2)$ on a rectangular coordinate system. Label this point as *A*.	**2.** Plot the point $(2,5)$ on the preceding rectangular coordinate system. Label this point as *B*.

Example

1. Plot the point $(5,2)$ on a rectangular coordinate system. Label this point as *A*.

Since the *x*-coordinate is 5, we first count 5 units to the right on the *x*-axis. Then, because the *y*-coordinate is 2, we count 2 units up from the point where we stopped on the *x*-axis. This locates the point corresponding to $(5,2)$. We mark this point with a dot and label it *A*.

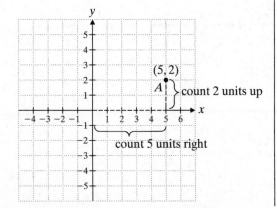

Vocabulary Answers: 1. coordinates 2. graphs 3. *y*-axis 4. *x*-coordinate

Example	Student Practice

3. Use the rectangular coordinate system to plot each point. Label the points F and G respectively.

(a) $(-5,-3)$

Notice that the x-coordinate, -5, is negative. On the coordinate grid, negative x-values appear to the left of the origin. Thus we will begin by counting 5 squares to the left, starting at the origin. Since the y-coordinate, -3, is negative, we will count 3 units down from the point where we stopped on the x-axis.

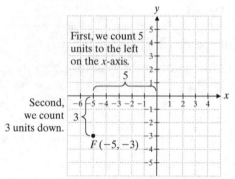

(b) $(2,-6)$

The x-coordinate is positive. Begin by counting 2 squares to the right of the origin. Then count down because the y-coordinate is negative.

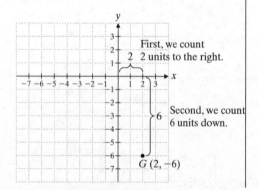

4. Use the rectangular coordinate system below to plot each point. Label the points I, J, and K, respectively.

(a) $(-1,-4)$

(b) $(-4,3)$

(c) $(3,-5)$

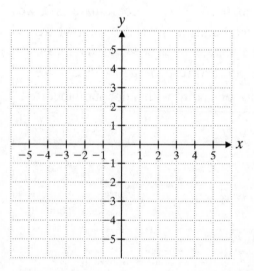

Example	Student Practice

5. What ordered pairs of numbers represent point A and point B on the graph?

To find point A, move along the x-axis until you get as close as possible to A, ending at 5. Thus obtaining 5 as the first number of the ordered pair. Then count 4 units upward on a line parallel to the y-axis to reach A. So you obtain 4 as the second number of the ordered pair. Thus point A is $(5,4)$. Use the same approach to find point B: $(-5,-3)$.

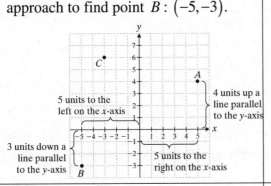

6. What ordered pair of numbers represents point C on the graph in Example **5**?

7. Is the ordered pair $(-1,4)$ a solution to the equation $3x+2y=5$?

We replace the values for x and y to see if we obtain a true statement.
Replace x with -1 and y with 4.

$$3x+2y=5$$
$$3(-1)+2(4)\overset{?}{=}5$$
$$-3+8\overset{?}{=}5$$
$$5=5$$

The ordered pair $(-1,4)$ is a solution to $3x+2y=5$ because when we replace x with -1 and y with 4, we obtain a true statement.

8. Is the ordered pair $(3,-3)$ a solution to the equation $3x+2y=5$?

201

Example	Student Practice
9. Find the missing coordinate to complete the ordered-pair solution $(0,?)$ to the equation $2x+3y=15$.	**10.** Find the missing coordinate to complete the ordered-pair solution $(?,3)$ to the equation $6x+5y=3$.

For the ordered pair $(0,?)$, we know that $x=0$. Replace x with 0 in the equation and solve for y.

$$2x+3y=15$$
$$2(0)+3y=15$$
$$0+3y=15$$
$$y=5$$

Thus we have the ordered pair $(0,5)$.

Extra Practice

1. Plot the point $D:(-1,-3)$.

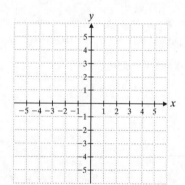

2. Consider the point plotted on the graph below. Give the coordinates for point A.

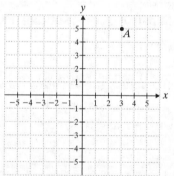

3. Find the missing coordinate to complete the ordered-pair solution to $y=-2x+3$.

(a) $(-1,?)$

(b) $(3,?)$

4. Find the missing coordinate to complete the ordered-pair solution to $4x-2y=12$.

(a) $(-1,?)$

(b) $(4,?)$

Concept Check

Explain how you would find the missing coordinate to complete the ordered-pair solution to the equation $2.5x+3y=12$ if the ordered pair was of the form $(?,-6)$.

Name: _____   Date: _____
Instructor: _____   Section: _____

Chapter 7 Graphing and Functions
7.2 Graphing Linear Equations

Vocabulary
linear equation   • x-intercept   • y-intercept   •   horizontal   •   vertical

1. The _____ of a line is the point where the line crosses the y-axis.

2. The graph of any _____ in two variables is a straight line.

3. The _____ of a line is the point where the line crosses the x-axis.

Example	Student Practice
1. Find three ordered pairs that satisfy $y = -2x + 4$. Then graph the resulting straight line.	**2.** Find three ordered pairs that satisfy $y = 2x - 3$. Then graph the resulting straight line. Use the given coordinate system.

Let $x = 0$, $x = 1$, and $x = 3$. For each x-value, find the corresponding y-value in the equation. Place each y-value in the table next to its x-value.

x	y
0	4
1	2
3	-2

If we plot these ordered pairs and connect the three points, we get a straight line that is the graph of the equation.

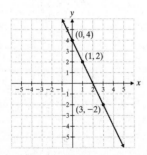

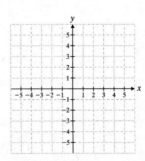

Vocabulary Answers: 1. y-intercept   2. linear equation   3. x-intercept

Example	Student Practice

Example

3. Graph $5x - 4y + 2 = 2$.

First, we simplify the equation by subtracting 2 from each side.

$$5x - 4y + 2 = 2$$
$$5x - 4y + 2 - 2 = 2 - 2$$
$$5x - 4y = 0$$

Since we are free to choose any value of x, $x = 0$ is a natural choice. Calculate the value of y when $x = 0$.

$$5(0) - 4y = 0$$
$$-4y = 0$$
$$y = 0$$

A convenient choice for a replacement of x is a number that is divisible by 4. Let $x = 4$ and $x = -4$. Follow the process used above to find the corresponding y-values and place the results in the table of values.

x	y
0	0
4	5
-4	-5

Graph the line.

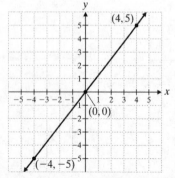

Student Practice

4. Graph $4x - 2y + 7 = 7$.

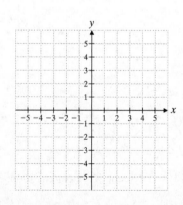

Example	Student Practice

Example

5. Complete **(a)** and **(b)** for the equation $5y - 3x = 15$.

   (a) State the x- and y-intercepts.

     Let $y = 0$.

$$5(0) - 3x = 15$$
$$-3x = 15$$
$$x = -5$$

     $(-5, 0)$ is the x-intercept. Now let $x = 0$.

$$5y - 3(0) = 15$$
$$5y = 15$$
$$y = 3$$

     $(0, 3)$ is the y-intercept.

   (b) Use the intercept method to graph.

     Find a third point and then graph. If we let $y = 6$, $x = 5$. The ordered pair is $(5, 6)$.

x	y
−5	0
0	3
5	6

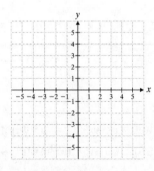

Student Practice

6. Complete **(a)** and **(b)** for the equation $y + 2x = -2$.

   (a) State the x- and y-intercepts.

   (b) Use the intercept method to graph.

Example	Student Practice
7. Graph $2x + 1 = 11$.	**8.** Graph $y + 4 = 7$.

Notice that there is only one variable, x, in the equation. Simplifying the equation yields $x = 5$. Since the x-coordinate of every point on this line is 5, we can see that the vertical line will be 5 units to the right of the y-axis.

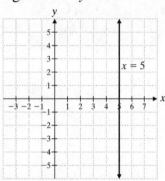

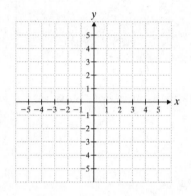

Extra Practice

1. Graph $y = 2x - 5$ by plotting three points and connecting them.

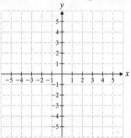

2. Graph $5x + 2y = 10$ by plotting three points and connecting them.

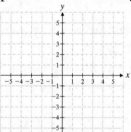

3. Graph $y = 5 - x$ by plotting intercepts and one other point.

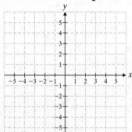

4. Graph $2x - 5 = y$ by plotting intercepts and one other point.

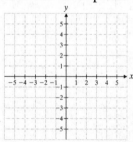

Concept Check

In graphing the equation $3y - 7x = 0$, what is the most important ordered pair to obtain before drawing a graph of the line? Why is that ordered pair so essential to drawing the graph?

206

Chapter 7 Graphing and Functions
7.3 The Slope of a Line

Vocabulary

slope   •   positive slope   •   negative slope   •   zero slope
undefined slope   •   parallel lines   •   perpendicular lines

1. _____ have slops whose product is -1.

2. _____ have the same slope but different y-intercepts.

3. In a coordinate plane, the _____ of a straight line is defined by the change in y divided by the change in x.

4. A vertical line is said to have _____.

Example	**Student Practice**
1. Find the slope of the line that passes through $(2,0)$ and $(4,2)$. Let $(2,0)$ be the first point (x_1, y_1) and $(4,2)$ be the second point (x_2, y_2). $$\text{slope} = m = \frac{y_2 - y_1}{x_2 - x_1} = \frac{2-0}{4-2} = \frac{2}{2} = 1$$ Note that the slope of the line will be the same if we let $(4,2)$ be the first point (x_1, y_1) and $(2,0)$ be the second point (x_2, y_2). $$m = \frac{y_2 - y_1}{x_2 - x_1} = \frac{0-2}{2-4} = \frac{-2}{-2} = 1$$ Thus, it does not matter which point you call (x_1, y_1) and which you call (x_2, y_2).	2. Find the slope of the line that passes through $(0,-4)$ and $(-3,-6)$.

Vocabulary Answers: 1. perpendicular lines 2. parallel lines 3. slope 4. undefined slope

Example	Student Practice
3. Find the slope of the line that passes through the given points. **(a)** $(0,2)$ and $(5,2)$ Calculate the slope. $$m = \frac{y_2 - y_1}{x_2 - x_1} = \frac{2-2}{5-0} = \frac{0}{5} = 0$$ The slope of a horizontal line is 0. **(b)** $(-4,0)$ and $(-4,-4)$ Calculate the slope. $$m = \frac{y_2 - y_1}{x_2 - x_1} = \frac{-4-0}{-4-(-4)} = \frac{-4}{0}$$ Recall that division by 0 is undefined. The slope of a vertical line is undefined.	**4.** Find the slope of the line that passes through the given points. **(a)** $(7,3)$ and $(7,-4)$ **(b)** $(5,3)$ and $(-1,3)$
5. What is the slope and the y-interept of the line $5x + 3y = 2$? We want to solve for y and get the equation in the form $y = mx + b$. $5x + 3y = 2$ $\quad 3y = -5x + 2$ $\qquad y = -\frac{5}{3}x + \frac{2}{3}$ $m = -\frac{5}{3}$ and $b = \frac{2}{3}$ The slope is $-\frac{5}{3}$. The y-interept is $\left(0, \frac{2}{3}\right)$.	**6.** What is the slope and the y-interept of the line $9x + 3y = 12$?

Example	Student Practice
7. Find an equation of the line with slope $\frac{2}{5}$ and y-interept $(0,-3)$. Write the equation in slope-intercept form, $y = mx + b$.	**8.** Find an equation of the line with slope $\frac{3}{4}$ and y-interept $(0,-7)$.

We are given that $m = \frac{2}{5}$ and $b = -3$.
Thus we have the following.

$$y = mx + b$$

$$y = \frac{2}{5}x + (-3)$$

$$y = \frac{2}{5}x - 3$$

(a) Write the equation in slope-intercept form, $y = mx + b$.

(b) Write the equation in the form $Ax + By = C$.

---

9. Graph the equation $y = -\frac{1}{2}x + 4$.

Begin with the y-intercept. Since $b = 4$, plot the point $(0,4)$. The slope, $-\frac{1}{2}$ can be written as $\frac{-1}{2}$. Begin at $(0,4)$ and go down 1 unit and to the right 2 units. This is the point $(2,3)$. Plot the point. Draw the line that connects the two points. This is the graph of the equation $y = -\frac{1}{2}x + 4$.

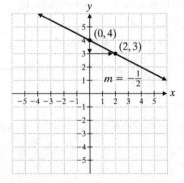

10. Graph the equation $y = -\frac{3}{4}x + 1$.

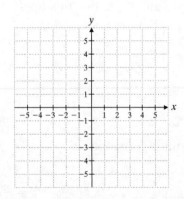

Example	Student Practice
11. Line h has a slope of $-\dfrac{2}{3}$.	**12.** Line h has a slope of $\dfrac{6}{7}$.

11. Line h has a slope of $-\dfrac{2}{3}$.

   (a) If line f is parallel to line h, what is its slope?

   Parallel lines have the same slope.

   Line f has a slope of $-\dfrac{2}{3}$.

   (b) If line g is perpendicular to line h, what is its slope?

   Perpendicular lines have slopes whose product is -1. So $m_1 m_2 = -1$.

   Substitute $-\dfrac{2}{3}$ for m_1.

$$m_1 m_2 = -1$$

$$-\frac{2}{3} m_2 = -1$$

$$\left(-\frac{3}{2}\right)\left(-\frac{2}{3}\right) m_2 = -1\left(-\frac{3}{2}\right)$$

$$m_2 = \frac{3}{2}$$

   Thus line g has a slope of $\dfrac{3}{2}$.

12. Line h has a slope of $\dfrac{6}{7}$.

   (a) If line f is parallel to line h, what is its slope?

   (b) If line g is perpendicular to line h, what is its slope?

Extra Practice

1. Find the slope of a straight line that passes through the points $(-5,-2)$ and $(1,-4)$.

2. Find the slope and the y-intercept of the line $y = 5x$.

3. Write the equation of the line in slope-intercept form given $m = -4$ and the y-intercept is $\left(0, \dfrac{4}{5}\right)$.

4. A line has a slope of 3.
   (a) What is the slope of a line parallel to it?
   (b) What is the slope of a line perpendicular to it?

Concept Check

Consider the formula for slope: $m = \dfrac{y_2 - y_1}{x_2 - x_1}$. Explain why we substitute the y coordinates in the numerator.

Chapter 7 Graphing and Functions
7.4 Writing the Equation of a Line

Vocabulary
slope • y-intercept • slope-intercept form • vertical units • horizontal units

1. The _____ is the change in y divided by the change in x.

2. If we know the slope and the y-intercept, we can write the equation of the line in _____.

3. Given the slope and a point on the line we can find the _____.

4. To find the slope of a line given the graph, we count the number of _____ and horizontal units from the one point on the line to another.

Example	Student Practice
1. Find an equation of the line that passes through $(-3,6)$ with slope $-\dfrac{2}{3}$.	**2.** Find an equation of the line that passes through $(2,-5)$ with slope $-\dfrac{1}{2}$.

We are given the values $m = -\dfrac{2}{3}$, $x = -3$, and $y = 6$. Substitute the given values of x, y, and m into the equation $y = mx + b$. Solve for b.

$y = mx + b$

$6 = \left(-\dfrac{2}{3}\right)(-3) + b$

$6 = 2 + b$

$4 = b$

Use the values of b and m to write the equation in the form $y = mx + b$.

An equation of the line is $y = -\dfrac{2}{3}x + 4$.

Vocabulary Answers: 1. slope 2. slope-intercept form 3. y-intercept 4. vertical units

Example	Student Practice
3. Find an equation of the line that passes through $(2,5)$ and $(6,3)$.	**4.** Find an equation of the line that passes through $(5,4)$ and $(10,1)$.

We first find the slope of the line. Then proceed as in Example **1**.

Substitute $(x_1, y_1) = (2,5)$ and $(x_2, y_2) = (6,3)$ into the formula.

$$m = \frac{y_2 - y_1}{x_2 - x_1}$$

$$m = \frac{y_2 - y_1}{x_2 - x_1} = \frac{3-5}{6-2} = \frac{-2}{4} = -\frac{1}{2}$$

Choose either point, say $(2,5)$, to substitute into $y = mx + b$ as in Example **1**. Then solve for b.

$$y = mx + b$$
$$5 = -\frac{1}{2}(2) + b$$
$$5 = -1 + b$$
$$6 = b$$

Use the values for b and m to write the equation.

An equation of the line is $y = -\frac{1}{2}x + 6$.

Note: We could have substituted the slope and the other point, $(6,3)$, into the slope-intercept form and arrived at the same answer. Try it.

Example	**Student Practice**
5. What is the equation of the line in the figure below?	**6.** What is the equation of the line in the figure below?

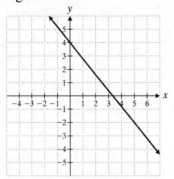

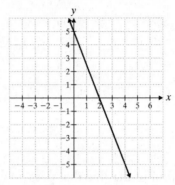

First, look for the y-intercept. The line crosses the y-axis at $(0,4)$. Thus $b = 4$. Second, find the slope.

$$m = \frac{\text{change in } y}{\text{change in } x}$$

Look for another point on the line. We choose $(5,-2)$. Count the number of vertical units from 4 to -2 (rise). Count the number of horizontal units from 0 to 5 (run).

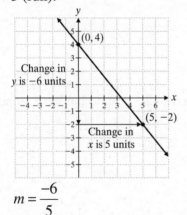

$$m = \frac{-6}{5}$$

Now, using $m = \dfrac{-6}{5}$ and $b = 4$, we can write an equation of the line.

$$y = mx + b$$

$$y = -\frac{6}{5}x + 4$$

Extra Practice

1. Find the equation of the line that passes through $(3,1)$ and has slope $-\dfrac{1}{2}$.

2. Write an equation of the line that passes through $(3,4)$ and $(-1,-16)$.

3. Write an equation of the line that passes through $\left(1,\dfrac{1}{6}\right)$ and $\left(2,\dfrac{4}{3}\right)$.

4. Find the equation of a line that passes through $(3,-7)$ and is parallel to $y=-5x+2$.

Concept Check

How would you find an equation of the line that passes through $(-2,-3)$ and has zero slope?

Chapter 7 Graphing and Functions
7.5 Graphing Linear Inequalities

Vocabulary
linear inequality • solution • solid line • dashed line • test point

1. The _____ of an inequality is the set of all possible ordered pairs that when substituted into the inequality will yield a true statement.

2. If the _____ is a solution of the inequality, we shade the region on the side of the line that includes the point.

3. We use a _____ to indicate that the points on the line are included in the solution of the inequality.

Example	**Student Practice**
1. Graph $5x + 3y > 15$.	**2.** Graph $4x + 3y > 12$.

Begin by graphing the line $5x + 3y = 15$. Since there is no equals sign in the inequality, draw a dashed line to indicate that the line is not part of the solution set. The easiest test point to test is $(0,0)$. Substitute $(0,0)$ for (x, y).

$$5x + 3y > 15$$

$$5(0) + 3(0) > 15$$

$$0 > 15 \quad \text{false}$$

$(0,0)$ is not a solution. Shade the region on the side of the line that does not include $(0,0)$.

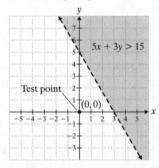

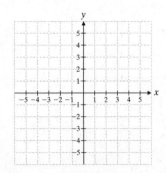

Vocabulary Answers: 1. solution 2. test point 3. solid line

Example	Student Practice

3. Graph $2y \le -3x$.

First, graph $2y = -3x$. Since $\le$ is used, the line will be a solid line.

We see that the line passes through $(0,0)$. We choose another test point. We will choose $(-3,-3)$.

$$2y \le -3x$$
$$2(-3) \le -3(-3)$$
$$-6 \le 9 \quad \text{true}$$

Since $(-3,-3)$ is a solution to the inequality, shade the region that includes $(-3,-3)$, that is the region below the line.

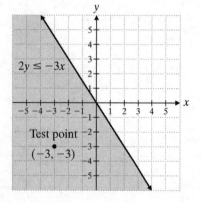

4. Graph $4y \le -5x$.

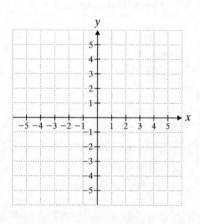

Example	**Student Practice**

5. Graph $x < -2$.

First, graph $x = -2$. Since $<$ is used, the line will be dashed.

Second, test $(0,0)$ in the inequality.

$x < -2$

$0 < -2$ false

Since $(0,0)$ is not a solution to the inequality, shade the region that does not include $(0,0)$, that is the region to the left of the line $x = -2$.

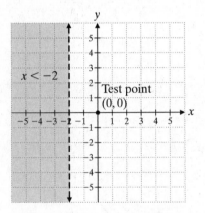

Observe that every point in the shaded region has an x-value that is less than -2.

6. Graph $x > -3$.

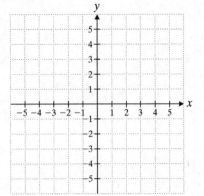

217

Extra Practice

1. Graph $y < 2x + 1$.

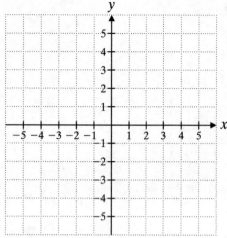

2. Graph $3x - 5y - 10 \geq 0$.

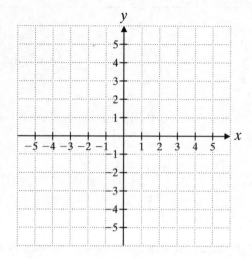

3. Graph $y \leq 4$.

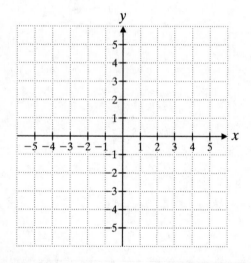

4. Graph $3x + 6y - 9 < 0$.

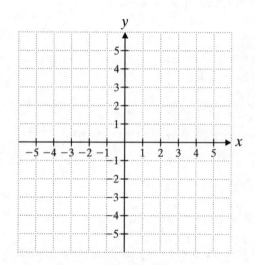

Concept Check

Explain how you would determine if you should shade the region above the line or below the line if you were to graph the inequality $y > -3x + 4$ using $(0,0)$ as a test point.

Chapter 7 Graphing and Functions
7.6 Functions

Vocabulary
independent variable • dependent variable • relation • domain • range
function • absolute zero • vertical line test • function notation

1. A(n) _____ is any set of ordered pairs.

2. A(n) _____ is a relation in which no two different ordered pairs have the same first coordinate.

3. The _____ is used to determine whether a relation is a function.

4. All the first coordinates in all of the ordered pairs of the relation make up the _____ of the relation.

Example	**Student Practice**
1. State the domain and range of the relation. $\{(5,7),(9,11),(10,7),(12,14)\}$	**2.** State the domain and range of the relation. $\{(-3,6),(7,1),(-2,1),(4,6)\}$

The domain consists of all the first coordinates in the ordered pairs. The first coordinates are 5, 9, 10, and 12.

The range consists of all the second coordinates in the ordered pairs. The second coordinates are 7, 11, 7, and 14.

We usually list the values of a domain or range from smallest to largest.

The domain is $\{5,9,10,12\}$.

The range is $\{7,11,14\}$.

Note that we list 7 only once.

Vocabulary Answers: 1. relation 2. function 3. vertical line test 4. domain

Example	Student Practice

3. Determine whether the relation is a function.

 (a) $\{(3,9),(4,16),(5,9),(6,36)\}$

 No two ordered pairs have the same first coordinate. Thus this set of ordered pairs defines a function.

 (b) $\{(7,8),(9,10),(12,13),(7,14)\}$

 Two different ordered pairs, $(7,8)$ and $(7,14)$ have the same first coordinate. Thus this relation is not a function.

4. Determine whether the relation is a function.

 (a) $\{(9,3),(16,4),(9,5),(36,6)\}$

 (b) $\{(-3,17),(2,1),(4,-3),(7,17)\}$

---

5. Graph $y = x^2$.

Begin by constructing a table of values. We select values for x and then determine by the equation the corresponding values of y. We then plot the ordered pairs and connect the points with a smooth curve.

x	$y = x^2$	y
-2	$y = (-2)^2 = 4$	4
-1	$y = (-1)^2 = 1$	1
0	$y = (0)^2 = 0$	0
1	$y = (1)^2 = 1$	1
2	$y = (2)^2 = 4$	4

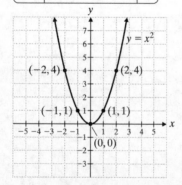

6. Graph $x = y^2$.

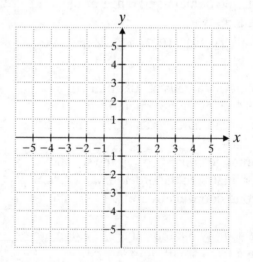

220

Example	Student Practice
7. Determine whether each of the following is the graph of a function.	**8.** Determine whether each of the following is the graph of a function.

(a)

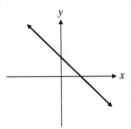

(a)

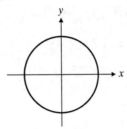

The graph of a straight line is a function. Any vertical line will cross this straight line in only one location.

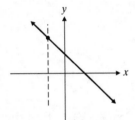

(b)

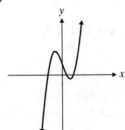

(b)

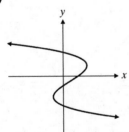

This graph is not the graph of a function. There exists a vertical line that will cross the curve in more than one place.

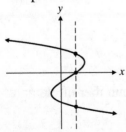

(c)

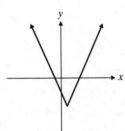

Example	Student Practice
9. If $f(x) = 3x^2 - 4x + 5$, find each of the following.	**10.** If $f(x) = 4x^2 - 2x + 7$, find each of the following.
(a) $f(-2)$	**(a)** $f(-3)$

$$f(-2) = 3(-2)^2 - 4(-2) + 5$$
$$= 3(4) + -4(-2) + 5$$
$$= 12 + 8 + 5 = 25$$

(b) $f(4)$	**(b)** $f(2)$

$$f(4) = 3(4)^2 - 4(4) + 5$$
$$= 3(16) + -4(4) + 5$$
$$= 48 - 16 + 5 = 37$$

(c) $f(0)$	**(c)** $f(0)$

$$f(0) = 3(0)^2 - 4(0) + 5 = 5$$

Extra Practice

1. Find the domain and range of the relation. Determine whether the relation is a function.

$$\{(2.5, 3), (3.5, 0), (5.5, -2), (8.5, -6)\}$$

2. Given $f(x) = 2x^2 - 3x + 1$, find the indicated values.

(a) $f(0)$

(b) $f(-3)$

(c) $f(3)$

3. Determine whether the relation is a function.

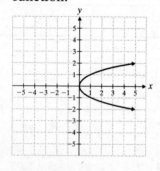

4. Graph $y = -3x^2$.

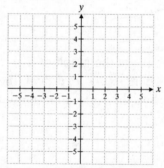

Concept Check

In the relation $\{(3,4), (5,6), (3,8), (2,9)\}$, why is there a different number of elements in the domain than the range?

MATH COACH

Mastering the skills you need to do well on the test.

Watch the MATH COACH videos in MyMathLab®or on YouTube™ while you work the problems below. These helpful hints will help you avoid making common errors on test problems.

Graphing a Linear Equation by Plotting Three Ordered

Pairs—Problem 8   Graph $y = \frac{2}{3}x - 4$.

Helpful Hint: Find three ordered pairs that are solutions to the equation. Plot those 3 points. Then draw a line through the points. When the equation is solved for y and there are fractional coefficients on x, it is sometimes a good idea to choose values for x that result in integer values for y. This will make graphing easier.

Choosing values for x that result in integer values for y means choosing 0 or multiples of 3 since 3 is the denominator of the fractional coefficient on x. When we multiply by these values, the result becomes an integer.

Did you choose values for x that result in values for y that are not fractions? Yes _____ No _____

If you answered No, try using $x = 0$, $x = 3$, and $x = 6$. Solve the equation for y in each case to find the y-coordinate. Remember that it will make the graphing process easier if you choose values for x that will clear the fraction from the equation.

Did you plot the three points and connect them with a line? Yes _____ No _____

If you answered No, go back and complete this step.

If you feel more comfortable using the slope-intercept method to solve this problem, simply identify the y-intercept from the equation in $y = mx + b$ form, and then use the slope m to find two other points.

If you answered Problem 4 incorrectly, go back and rework the problem using these suggestions.

Write the Equation of a Line Given Two Points—Problem 10

Find an equation for the line passing through $(5, -4)$ and $(-3, 8)$.

Helpful Hint: When given two points (x_1, y_1) and (x_2, y_2), we can find the slope m using the slope formula $m = \frac{y_2 - y_1}{x_2 - x_1}$. Then you can substitute m into the equation $y = mx + b$ along with the coordinates of one of the points to find b, the y-intercept.

When you substituted the points into the slope formula to find m, did you obtain either $m = \frac{8 - (-4)}{-3 - 5}$ or

$m = \frac{-4 - 8}{5 - (-3)}$?

Yes _____ No _____

If you answered No, check your work to make sure you substituted the points correctly. Be careful to avoid any sign errors.

(continued on next page)

Did you use $m = -\dfrac{3}{2}$ and either of the points given when you substituted the values into the equation $y = mx + b$ to find the value of b? Yes _____ No _____

If you answered No, stop and make a careful substitution for $m = -\dfrac{3}{2}$ and either $x = 5$ and $y = -4$ or $x = -3$ and $y = 8$.

See if you can solve the resulting equation for b.

Now go back and rework the problem using these suggestions.

Graphing Linear Inequalities in Two Variables—Problem 12

Graph the region described by $-3x - 2y > 10$.

Helpful Hint: First graph the equation $-3x - 2y = 10$. Determine if the line should be solid or dashed. Then pick a test point to see if it satisfies the inequality $-3x - 2y > 10$. If the test point satisfies the inequality, shade the side of the line on which the point lies. If the test point does not satisfy the inequality, shade the opposite side of the line.

Examine your work. Does the line $-3x - 2y = 10$ pass through the point $(0, -5)$? Yes _____ No _____

If you answered No, substitute $x = 0$ into the equation and solve for y. Check the calculations for each of the points you plotted to find the graph of this equation.

Did you draw a solid line? Yes _____ No _____

If you answered Yes, look at the inequality symbol. Remember that we only use a solid line with the symbols $\leq$ and $\geq$. A dashed line is used for $<$ and $>$.

Did you shade the area above the dashed line? Yes _____ No _____

If you answered Yes, stop now and use $(0, 0)$ as a test point and substitute it into the inequality $-3x - 2y > 10$. Then use the Helpful Hint to determine which side to shade.

If you answered Problem 12 incorrectly, go back and rework the problem using these suggestions.

Using Function Notation to Evaluate a Function—Problem 16(a) and 16(b)

For $f(x) = -x^2 - 2x - 3$:   **(a)** find $f(0)$.   **(b)** find $f(-2)$.

Helpful Hint: Replace x with the number indicated. It is a good idea to place parentheses around the value to avoid any sign errors. Then use the order of operations to evaluate the function in each case.

(a) Did you replace x with 0 and write
$$f(0) = -(0)^2 - 2(0) - 3?$$
Yes _____ No _____

If you answered No, take time to go over your steps one more time, remembering that 0 times any number is 0.

(b) Did you replace x with -2 and write
$$f(0) = -(0)^2 - 2(0) - 3?$$
Yes _____ No _____

If you answered No, go over your steps again, remembering to place parentheses around -2.

Note that $(-2)^2 = 4$ and therefore
$$-(-2)^2 = -4.$$

Now go back and rework the problem again using these suggestions.

Name: _____   Date: _____
Instructor: _____   Section: _____

Chapter 8 Systems of Equations
8.1 Solving a System of Equations in Two Variables by Graphing

Vocabulary
system of equations • consistent • inconsistent • dependent • coordinate system
slope • linear equation • intersection

1. A linear system of equations that has one solution is said to be _____.

2. A(n) _____, is any set of equations in multiple variables that is considered at the same time.

3. A system of equations with infinite solutions is called _____.

4. A system of equations that has no solution is said to be _____.

Example	**Student Practice**
1. Solve by graphing. $\begin{array}{c} -2x + 3y = 6 \\ 2x + 3y = 18 \end{array}$	2. Solve by graphing. $\begin{array}{c} 2x + 2y = 4 \\ -3x + y = -2 \end{array}$

Solve each equation for y. Use a table to find and graph a series of ordered pairs.

$$y = \frac{2}{3}x + 2$$

$$y = -\frac{2}{3}x + 6$$

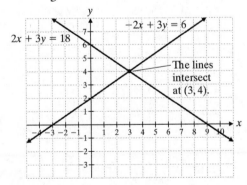

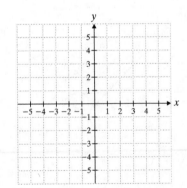

The unique solution is the point of intersection, $(3, 4)$.

Vocabulary Answers: 1. consistent 2. system of equations 3. dependent 4. inconsistent

Example	Student Practice

Example

3. Solve by graphing. $\begin{aligned} 3x - y &= 1 \\ 3x - y &= -7 \end{aligned}$

Graph both equations on the same coordinate system.

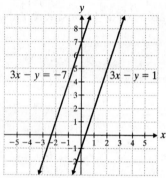

The two lines are parallel. A system of linear equations that does not intersect does not have a solutions and is called inconsistent.

5. Solve by graphing. $\begin{aligned} x + y &= 4 \\ 3x + 3y &= 12 \end{aligned}$

Graph both equations on the same coordinate system.

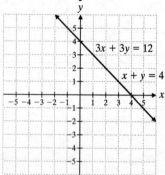

The two equations represent the same line; they coincide. There is an infinite number of solutions to this system. Such equations are said to be dependent.

Student Practice

4. Solve by graphing. $\begin{aligned} 2x + 4y &= 0 \\ \frac{1}{2}x + y &= 3 \end{aligned}$

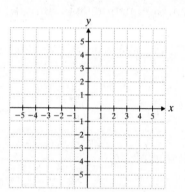

6. Solve by graphing. $\begin{aligned} -3x + 6y &= 24 \\ -x + 2y &= 8 \end{aligned}$

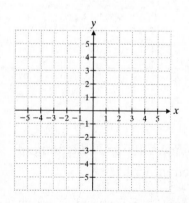

226

Example	Student Practice

Example

7. Roberts Plumbing and Heating charges $40 for a house call and then $35 per hour for labor. Instant Plumbing Repairs charges $70 for a house call and then $25 per hour for labor.

(a) Create a cost equation for each company, where y is the total cost of plumbing repairs and x is the number of hours per labor. Write a system of equations.

For each company obtain a cost equation.

$$\begin{array}{ccccc} \textit{Cost of} & \textit{house} & \textit{per} & \textit{labor} \\ \textit{plumbing} = \textit{call} & + & \textit{hour} \times \textit{hours} \end{array}$$

$$y = 40 + 35 \times x$$

$$y = 70 + 25 \times x$$

(b) Graph the two equations and determine from your graph how many hours of plumbing repairs would be required for the two companies to charge the same.

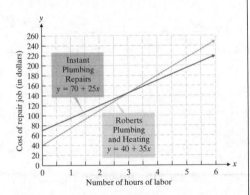

We see that the graphs of the two lines intersect at $(3, 145)$. Thus the two companies will charge the same if 3 hours of plumbing repairs are required.

Student Practice

8. Suppose Walter and Barbara try two more plumbers. Plumbers A charges $50 for a house call and then $40 per hour for labor. Plumbers B charges $90 for a house call and then $30 per hour for labor.

(a) Create a cost equation for each company, where y is the total cost of plumbing repairs and x is the number of hours per labor. Write a system of equations.

(b) Graph the two equations.

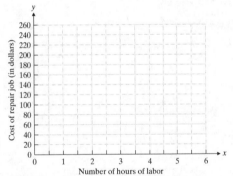

(c) Determine from your graph how many hours of plumbing repairs would be required for the two companies to charge the same.

(d) Determine from your graph which company charges less if the estimated amount of time to complete the plumbing repairs is 5 hours.

227

Extra Practice

1. Solve by graphing. If there isn't a unique solution to the system, state the reason.
$$3x - y = 0$$
$$2x + y = -10$$

2. Solve by graphing. If there isn't a unique solution to the system, state the reason.
$$-6x + 2y = 5$$
$$3x - y = 3$$

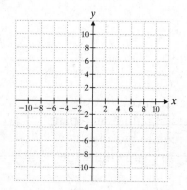

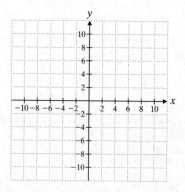

3. Solve by graphing. If there isn't a unique solution to the system, state the reason.
$$2x + 3y = 18$$
$$3x - 2y = 14$$

4. Solve by graphing. If there isn't a unique solution to the system, state the reason.
$$4x - 8y = 20$$
$$-2x + 4y = -10$$

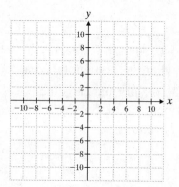

Concept Check

You are attempting to find the solution to this system of equations by graphing.

$$-3x + 4y = -16$$
$$0 = 6x - 8y + 16$$

What will you discover?

Chapter 8 Systems of Equations
8.2 Solving a System of Equations in Two Variables by the Substitution Method

Vocabulary

substitution method   •   system of equations   •   consistent   •   inconsistent

linear equation   •   dependent

1. An equation of the form $Ax + By = C$ is called a(n) _____.

2. The _____ is a strategy for solving a system of two linear equations that involves reducing the system to one equation in one variable which can be solved for.

Example	Student Practice
1. Find the solution. $x - 2y = 7$ (1) $-5x + 4y = -5$ (2)	**2.** Find the solution. $3x + y = -3$ $-5x - 2y = 4$

Solve equation (1) for x.

$x = 7 + 2y$

Substitute this expression into (2).

$-5x + 4y = -5$

$-5(7 + 2y) + 4y = -5$

Solve this equation for y.

$-35 - 10y + 4y = -5$

$-6y = 30$

$y = -5$

Use the value for y to find the value for x.

$x = 7 + 2y$

$x = 7 + 2(-5)$

$x = -3$

The solution is $(-3, -5)$.

Vocabulary Answers: 1. linear equation 2. substitution method

Example	Student Practice
3. Find the solution. $\dfrac{3}{2}x + y = \dfrac{5}{2}$ (1) $-y + 2x = -1$ (2)	**4.** Find the solution. $\dfrac{4}{3}x - \dfrac{2}{3}y = \dfrac{1}{12}$ $-y - x = -1$

Clear the first equation of fractions.

$$2\left(\frac{3}{2}x\right) + 2(y) = 2\left(\frac{5}{2}\right)$$

$$3x + 2y = 5$$

The new system is as follows.

$$3x + 2y = 5 \quad (3)$$
$$-y + 2x = -1 \quad (2)$$

Solve for y in equation (2).

$$y = 1 + 2x$$

Substitute this expression into (3) and solve for x.

$$3x + 2(1 + 2x) = 5$$
$$7x + 2 = 5$$
$$x = \frac{3}{7}$$

Use the value for x to find the value for y.

$$y = 1 + 2x$$
$$y = 1 + 2\left(\frac{3}{7}\right)$$
$$y = \frac{13}{7}$$

The solution is $\left(\dfrac{3}{7}, \dfrac{13}{7}\right)$.

Extra Practice

1. Find the solution to the system of equations by the substitution method. Write your solution in the form (a, b).

$$5a + 7b = -21$$
$$a + 18 = -6b$$

2. Find the solution to the system of equations by the substitution method. Write your solution in the form (x, y).

$$x = 6 + 4y$$
$$-3(x + y) = -36$$

3. Find the solution to the system of equations by the substitution method. Write your solution in the form (x, y).

$$8x - 5y = -78$$
$$6x - 3y = -54$$

4. Find the solution to the system of equations by the substitution method. Write your solution in the form (x, y).

$$5x + 8(y - 3) = 15$$
$$2(x + 2y) = 42$$

Concept Check

Explain how you would solve the following system by the substitution method.

$$3x - y = -4$$
$$-2x + 4y = 36$$

Name: _____   Date: _____
Instructor: _____   Section: _____

Chapter 8 Systems of Equations
8.3 Solving a System of Equations in Two Variables by the Addition Method

Vocabulary
substitution method • system of equations • addition method • elimination method

1. The addition method is often referred to as the _____.

2. The _____ is a strategy for solving a system of two linear equations that works well for integer, fractional, and decimal coefficients.

Example

1. Solve by addition. $5x + 2y = 7$ $\quad(1)$
$\qquad\qquad\qquad\quad 3x - y = 13$ $\quad(2)$

Make the coefficients of the y-terms opposites by multiplying each term of equation (2) by 2.

$2(3x) - 2(y) = 2(13)$

Multiply, then add the equations.

$5x + 2y = 7$
$\underline{6x - 2y = 26}$
$11x \quad\;\; = 33$
$\qquad x = 3$

Substitute $x = 3$ into (1) and solve for y.

$5(3) + 2y = 7$
$\quad 15 + 2y = 7$
$\qquad\quad 2y = -8$
$\qquad\quad y = -4$

The solution is $(3, -4)$.

Student Practice

2. Solve by addition. $\quad 5x - y = 21$
$\qquad\qquad\qquad\qquad 3x + 3y = 9$

Vocabulary Answers: 1. elimination method 2. addition method

Copyright © 2013 Pearson Education, Inc.

Example	Student Practice

Example

3. Solve. $x - \dfrac{5}{2}y = \dfrac{5}{2}$ (1)

$\qquad \dfrac{4}{3}x + y = \dfrac{23}{3}$ (2)

Multiply each term of (1) by 2 and each term of (2) by 3 to clear the fractions.

$$2(x) - 2\left(\dfrac{5}{2}y\right) = 2\left(\dfrac{5}{2}\right)$$

$$3\left(\dfrac{4}{3}x\right) + 3(y) = 3\left(\dfrac{22}{3}\right)$$

$$\begin{aligned} 2x - 5y &= 5 \qquad (3) \\ 4x + 3y &= 23 \qquad (4) \end{aligned}$$

Make the coefficients of the x-terms opposites by multiplying each term of equation (3) by -2. Add and solve for y.

$$\begin{aligned} -4x + 10y &= -10 \\ \underline{4x + 3y} &= \underline{23} \\ 13y &= 13 \\ y &= 1 \end{aligned}$$

Substitute $y = 1$ into (4) and solve for x.

$$\begin{aligned} 4x + 3(1) &= 23 \\ 4x &= 20 \\ x &= 5 \end{aligned}$$

The solution is $(5,1)$.

The check is left to the student.

Student Practice

4. Solve. $\dfrac{3}{4}x - \dfrac{5}{4}y = -\dfrac{9}{4}$

$\qquad \dfrac{1}{3}x + y = \dfrac{11}{3}$

Example	Student Practice

Example

5. Solve. $0.12x + 0.05y = -0.02$ (1)

$\quad\quad\quad\; 0.08x - 0.03y = -0.14$ (2)

Since the decimals are hundredths, multiply each term of both equations by 100.

$$100(0.12x) + 100(0.05y) = 100(-0.02)$$
$$100(0.08x) - 100(0.03y) = 100(-0.14)$$

$\quad 12x + 5y = -2 \quad\quad (3)$

$\quad\;\; 8x - 3y = -14 \quad\; (4)$

Make the coefficients of the y-terms opposites by multiplying each term of (3) by 3 and each term of (4) by 5.

$\quad 36x + 15y = -6$

$\underline{\quad 40x - 15y = -70}$

$\quad\quad\quad 76x = -76$

$\quad\quad\quad\;\; x = -1$

Substitute x into (3) and solve for y.

$$12(-1) + 5y = -2$$
$$y = 2$$

The solution is $(-1, 2)$.

To check, substitute the solution into the original equations.

$$0.12(-1) + 0.05(2) = -0.02$$
$$-0.02 = -0.02 \quad \text{Checks}$$
$$0.08(-1) - 0.03(2) = -0.14$$
$$-0.14 = -0.14 \quad \text{Checks}$$

Student Practice

6. Solve. $0.12x - 0.67y = -1.03$

$\quad\quad\quad\; 0.50x + 0.75y = -0.75$

Extra Practice

1. Find the solution by the addition method. Check your answers.

$$x - y = 6$$
$$7x - 9y = -4$$

2. Find the solution by the addition method. Check your answers.

$$7x + 13 = 9y$$
$$3x + 5y = -41$$

3. Find the solution by the addition method. Check your answers.

$$x + \frac{3}{2}y = 3$$
$$7x - 9y = -5$$

4. Find the solution by the addition method. Check your answers.

$$6x - y = 5$$
$$6x + 3y = -3$$

Concept Check

Explain how you would obtain an equivalent system that does not have decimals if you wanted to solve the following system.

$$0.5x + 0.4y = -3.4$$
$$0.02x + 0.03y = -0.08$$

Chapter 8 Systems of Equations
8.4 Review of Methods for Solving Systems of Equations

Vocabulary
substitution method   •   addition method   •   inconsistent   •   identity   •   dependent

1. The equation $7 = 7$ is always true; it is an example of a(n) _____.

2. The _____ works well if one or more variables have a coefficient of 1 or -1.

3. A(n) _____ system of linear equations is a system of parallel lines that are not equal.

4. _____ equations produce a system with an unlimited number of solutions.

Example

1. Select a method and solve the system of equations.

$$x + y = 3080$$
$$2x + 3y = 8740$$

Use substitution since there are x- and y-values that have coefficients of 1. Solve the first equation for y and substitute it into the other to find x.

$$y = 3080 - x$$
$$2x + 3(3080 - x) = 8740$$
$$2x + 9240 - 3x = 8740$$
$$x = 500$$

Substitute 500 for x and solve for y.

$$y = 3080 - 500$$
$$y = 2580$$

The solution is $(500, 2580)$.

Student Practice

2. Select a method and solve the system of equations.

$$3x - 4y = 13$$
$$9x - 2y = 119$$

Vocabulary Answers: 1. identity 2. substitution method 3. inconsistent 4. dependent

Example	Student Practice

3. Solve algebraically. $\begin{array}{l} 3x - y = -1 \\ 3x - y = -7 \end{array}$

4. Solve algebraically. $\begin{array}{l} -3x - y = -5 \\ 6x + 2y = -1 \end{array}$

The addition method would be very convenient in this case.

Multiply each term in the second equation by -1 and then add the two equations.

$$3x - y = -1$$
$$\underline{-3x + y = +7}$$
$$0 = 6$$

The statement $0 = 6$ is not true, thus there is no solution to this system of equations. The lines are parallel and the system is inconsistent.

---

5. Solve algebraically. $\begin{array}{l} x + y = 4 \\ 3x + 3y = 12 \end{array}$

6. Solve algebraically. $\begin{array}{l} 4x - y = 3 \\ 12x - 3y = 9 \end{array}$

Let us use the substitution method. Solve the first equation for y and substitute it into the second equation.

$$y = 4 - x$$
$$3x + 3(4 - x) = 12$$
$$3x + 12 - 3x = 12$$
$$12 = 12$$

$12 = 12$ is always true; it is an identity. This means all the solutions of one equation of the system are also solutions of the other equation. Therefore, the lines coincide and there are an infinite number of solutions to this system of equations.

Extra Practice

1. If possible, solve by an algebraic method. Otherwise, state that the problem has no solution or an infinite number of solutions.

$$9x + 1 = 7y$$

$$-2x - 4y = 28$$

2. If possible, solve by an algebraic method. Otherwise, state that the problem has no solution or an infinite number of solutions.

$$9x - 3y = 10$$

$$45x - 15y = 20$$

3. If possible, solve by an algebraic method. Otherwise, state that the problem has no solution or an infinite number of solutions.

$$\frac{3}{2}x + \frac{1}{2}y - 7 = 0$$

$$\frac{2}{5}x - \frac{3}{5}y + \frac{9}{5} = 0$$

4. If possible, solve by an algebraic method. Otherwise, state that the problem has no solution or an infinite number of solutions.

$$3x + y = -14$$

$$y = 3x + 4$$

Concept Check

Explain how you would remove the fractions in order to solve the following system.

$$\frac{3}{10}x + \frac{2}{5}y = \frac{1}{2},$$

$$\frac{1}{8}x - \frac{3}{16}y = -\frac{1}{2}$$

Chapter 8 Systems of Equations
8.5 Solving Word Problems Using Systems of Equations

Vocabulary
substitution method • addition method • elimination method • system of equations

1. When solving a word problem with two equations and two unknown it is often helpful to set up a(n) _____.

2. If one or more variables have a coefficient of 1 or -1, the _____ works well.

Example	**Student Practice**
1. A worker in a large post office is trying to verify the rate at which two electronic card-sorting machines operate. Yesterday the first machine sorted for 3 minutes and the second machine sorted for 4 minutes. The total workload both machines processed during that time was 10,300 cards. Two days ago the first machine sorted for 2 minutes and the second machine sorted for 3 minutes. The total workload both machines processed during that time period was 7400 cards. Can you determine the number of cards per minute sorted by each machine?	**2.** A landscape company is trying to verify the rate at which two employees can plant bulbs in a large park. Yesterday the first employee planted for 5 hours and the second employee planted for 2 hours. The total number of bulbs planted during that time was 115. Last week the first employee planted for 3 hours and the second employee planted for 4 hours. The total number of bulbs planted during that time was 125. Can you determine the number of bulbs planted per hour by each employee?
Let $x =$ the number of cards per minute sorted by the first machine and $y =$ the number of cards per minute sorted by the second machine and write a system of two equations with two unknowns using the information from the two days.	
$3x + 4y = 10,300$ Yesterday	
$2x + 3y = 7400$ Two days ago	
Solve the system using the addition method. The first machine sorts 1300 cards per minute and the second machine sorts 1600 cards per minute.	

Vocabulary Answers: 1. system of equations 2. substitution method

Example	Student Practice
3. A lab technician is required to prepare 200 liters of a solution. The prepared solution must contain 42% fungicide. The technician wishes to combine a solution that contains 30% fungicide with a solution that contains 50% fungicide. How much of each solution should he use?	**4.** A chemistry student is required to prepare 150 milliliters of a solution. The prepared solution must contain 15% alcohol. The student wishes to combine a solution that contains 25% alcohol with a solution that contains 10% alcohol. How much of each solution should he use?

We need to find the amount of each solution that will give us 200 liters of a 42% solution. Let $x =$ the number of liters of the 30% solution needed and $y =$ the number of liters of the 50% solution needed.

Write an equation for combining x and y to get 200 liters.

$$x + y = 200$$

Write an equation for the percent of fungicide in each solution (30% of x liters, 50% of y liters, and 42% of 200 liters).

$$0.3x + 0.5y = 0.42(200) = 84$$

Solve the following system of equations using the addition method.

$$x + y = 200$$
$$0.3x + 0.5y = 84$$

The technician should use 80 liters of the 30% fungicide and 120 liters of the 50% fungicide to obtain the required solution.

The check is left to the student.

Example	Student Practice
5. Mike recently rode his boat on Lazy River. He took a 48-mile trip up the river against the current in exactly 3 hours. He refueled and made the return trip in exactly 2 hours. What was the speed of his boat in still water and the speed of the current in the river?	**6.** Gary flew his small airplane 150 nautical miles to visit a friend. On the way there, he flew against a strong wind for 2.5 hours. On the way home, he flew in the same direction as the wind and made the trip in exactly 1.5 hours. What was the speed of his plane in still air and the speed of the wind that day?

Let $b =$ the speed of the boat in still water in miles/hour and let $c =$ the speed of the river current in miles/hour.

When we travel against the current, the current is slowing us down. Since the current's speed opposes the boat's speed in still water, we must subtract: $b - c$.

When we travel with the current, the current is helping us travel forward. The current's speed is added to the boat's speed in still water, we must add: $b + c$.

Use the distance formula,

$$distance = rate \times time$$

to write a system of equations.

$48 = (b - c) \cdot 3$ Against the current
$48 = (b + c) \cdot 2$ With the current

Remove the parenthesis to get the following system of equations.

$48 = 3b - 3c$ Against the current
$48 = 2b + 2c$ With the current

Solve the system of equations using the addition method to find that the speed of Mike's boat in still water was 20 miles/hour and the speed of the current in Lazy River was 4 miles/hour.

Chapter 9 Radicals
9.4 Multiplying Radical Expressions

Vocabulary
FOIL method   •   multiplication

1. The basic rule for _____ of square root radicals is $\sqrt{a}\sqrt{b} = \sqrt{ab}$.

2. The _____ can be used for binomial radical expressions.

Example	Student Practice
1. Multiply. $\sqrt{7}\sqrt{14x}$	**2.** Multiply. $\sqrt{10x}\sqrt{15}$
$\sqrt{7}\sqrt{14x} = \sqrt{98x}$	
We do not stop here, because the radical $\sqrt{98x}$ can be simplified.	
$\sqrt{98x} = \sqrt{49 \cdot 2x} = \sqrt{49}\sqrt{2x} = 7\sqrt{2x}$	
3. Multiply.	**4.** Multiply.
(a) $\left(2\sqrt{3}\right)\left(5\sqrt{7}\right)$	**(a)** $\left(4\sqrt{2}\right)\left(5\sqrt{3}\right)$
Multiply the coefficients: $2 \cdot 5 = 10$. Multiply the radicals: $\sqrt{3}\sqrt{7} = \sqrt{21}$.	
$\left(2\sqrt{3}\right)\left(5\sqrt{7}\right) = 10\sqrt{21}$	
	(b) $\left(5\sqrt{15z}\right)\left(9\sqrt{35z}\right)$
(b) $\left(2a\sqrt{3a}\right)\left(3\sqrt{6a}\right)$	
$\left(2a\sqrt{3a}\right)\left(3\sqrt{6a}\right) = 6a\sqrt{18a^2}$	
$= 6a\sqrt{9}\sqrt{2}\sqrt{a^2}$	
$= 18a^2\sqrt{2}$	

Vocabulary Answers: 1. multiplication 2. FOIL method

Example	Student Practice
5. Find the area of a farm field that measures $\sqrt{9400}$ feet long and $\sqrt{4800}$ feet wide. Express your answer as a simplified radical. We multiply. $\left(\sqrt{9400}\right)\left(\sqrt{4800}\right)=\left(10\sqrt{94}\right)\left(40\sqrt{3}\right)$ $=400\sqrt{282}$ square feet $\left(\text{ft}^2\right)$	**6.** Find the area of a swimming pool that measures $\sqrt{1200}$ feet long and $\sqrt{975}$ feet wide. Express your answer as a simplified radical.
7. Multiply and simplify. $\sqrt{5}\left(\sqrt{2}+3\sqrt{7}\right)$ $\sqrt{5}\left(\sqrt{2}+3\sqrt{7}\right)=\sqrt{5}\sqrt{2}+3\sqrt{5}\sqrt{7}$ $\qquad\qquad = \sqrt{10}+3\sqrt{35}$ In similar fashion, we can multiply a trinomial by another factor.	**8.** Multiply and simplify. $\sqrt{2}\left(3\sqrt{6}+7\sqrt{10}\right)$
9. Multiply. $\sqrt{a}\left(3\sqrt{a}-2\sqrt{5}\right)$ For any nonnegative real number a, $\sqrt{a}\cdot\sqrt{a}=a$. $\sqrt{a}\left(3\sqrt{a}-2\sqrt{5}\right)=3\sqrt{a}\sqrt{a}-2\sqrt{5}\sqrt{a}$ $=3a-2\sqrt{5a}$	**10.** Multiply. $3\sqrt{x}\left(x\sqrt{5}+7\sqrt{x}\right)$
11. Multiply. $\left(\sqrt{2}+5\right)\left(\sqrt{2}-3\right)$ $\left(\sqrt{2}+5\right)\left(\sqrt{2}-3\right)$ $=\sqrt{4}-3\sqrt{2}+5\sqrt{2}-15$ $=2+2\sqrt{2}-15$ $=-13+2\sqrt{2}$	**12.** Multiply. $\left(\sqrt{5}+3\right)\left(\sqrt{5}-2\right)$

Example	Student Practice
13. Multiply. $\left(\sqrt{2}-3\sqrt{6}\right)\left(\sqrt{2}+\sqrt{6}\right)$	**14.** Multiply. $\left(\sqrt{5}+\sqrt{15}\right)\left(2\sqrt{5}-6\sqrt{15}\right)$

$$\left(\sqrt{2}-3\sqrt{6}\right)\left(\sqrt{2}+\sqrt{6}\right)$$
$$=\sqrt{4}+\sqrt{12}-3\sqrt{12}-3\sqrt{36}$$
$$=2-2\sqrt{12}-18$$
$$=-16-4\sqrt{3}$$

15. Multiply. $\left(2\sqrt{3}-\sqrt{6}\right)^2$	**16.** Multiply. $\left(4\sqrt{7}-\sqrt{14}\right)^2$

$$\left(2\sqrt{3}-\sqrt{6}\right)^2$$
$$=\left(2\sqrt{3}-\sqrt{6}\right)\left(2\sqrt{3}-\sqrt{6}\right)$$
$$=4\sqrt{9}-2\sqrt{18}-2\sqrt{18}+\sqrt{36}$$
$$=12-4\sqrt{18}+6$$
$$=18-12\sqrt{2}$$

Extra Practice

1. Multiply. Be sure to simplify any radicals in your answer. Do not use a calculator or a table of square roots. $\sqrt{6}\sqrt{10}$

2. Multiply. Be sure to simplify any radicals in your answer. Do not use a calculator or a table of square roots. $\sqrt{5}\left(\sqrt{10}+\sqrt{x}\right)$

3. Multiply. Be sure to simplify any radicals in your answer. Do not use a calculator or a table of square roots.
$\left(5-2\sqrt{3}\right)\left(2+3\sqrt{6}\right)$

4. Multiply. Be sure to simplify any radicals in your answer. Do not use a calculator or a table of square roots.
$\left(2\sqrt{3x}-7\right)\left(2\sqrt{3x}+7\right)$

Concept Check

Explain how to simplify $\left(3\sqrt{3}-2\right)^2$.

Name: _____   Date: _____
Instructor: _____   Section: _____

Chapter 9 Radicals
9.5 Dividing Radical Expressions

Vocabulary
rationalize the denominator   •   conjugates   •   quotient rule

1. The _____ for square roots can be used to divide square root radicals or to simplify square root radical expressions involving division.

2. Expressions like $5 - 3\sqrt{2}$ and $5 + 3\sqrt{2}$ are called _____.

3. If a fraction has a radical in the denominator, we _____ by multiplying to change the fraction to an equivalent one that has an integer in the denominator.

Example	Student Practice
1. Simplify.	**2.** Simplify.
(a) $\dfrac{\sqrt{75}}{\sqrt{3}}$	(a) $\dfrac{\sqrt{180}}{\sqrt{5}}$
3 is a factor of 75. Use the quotient rule to rewrite the expression.	
$\dfrac{\sqrt{75}}{\sqrt{3}} = \sqrt{\dfrac{75}{3}}$	
$= \sqrt{25}$	
$= 5$	
	(b) $\sqrt{\dfrac{4}{9}}$
(b) $\sqrt{\dfrac{25}{36}}$	
Both 25 and 36 are perfect squares, so rewrite the expression.	
$\sqrt{\dfrac{25}{36}} = \dfrac{\sqrt{25}}{\sqrt{36}}$	
$= \dfrac{5}{6}$	

Vocabulary Answers: 1. quotient rule 2. conjugates 3. rationalize the denominator

Example	Student Practice
3. Simplify. $\sqrt{\dfrac{20}{x^6}}$	**4.** Simplify. $\sqrt{\dfrac{54}{y^2}}$

$$\sqrt{\frac{20}{x^6}} = \frac{\sqrt{20}}{\sqrt{x^6}} = \frac{\sqrt{4}\sqrt{5}}{x^3} = \frac{2\sqrt{5}}{x^3}$$

---

5. Simplify. $\dfrac{3}{\sqrt{2}}$

Think, "What times 2 will make a perfect square?"

$$\frac{3}{\sqrt{2}} = \frac{3}{\sqrt{2}} \times 1 = \frac{3}{\sqrt{2}} \times \frac{\sqrt{2}}{\sqrt{2}} = \frac{3\sqrt{2}}{\sqrt{4}} = \frac{3\sqrt{2}}{2}$$

6. Simplify. $\dfrac{7}{\sqrt{5}}$

---

7. Simplify.

   (a) $\dfrac{\sqrt{7}}{\sqrt{8}}$

      Think, "What times 8 will make a perfect square?"

$$\frac{\sqrt{7}}{\sqrt{8}} = \frac{\sqrt{7}}{\sqrt{8}} \times \frac{\sqrt{2}}{\sqrt{2}} = \frac{\sqrt{14}}{\sqrt{16}} = \frac{\sqrt{14}}{4}$$

   (b) $\dfrac{3}{\sqrt{x^3}}$

      Think, "What times x^3 will give an even exponent?"

$$\frac{3}{\sqrt{x^3}} \times \frac{\sqrt{x}}{\sqrt{x}} = \frac{3\sqrt{x}}{\sqrt{x^4}} = \frac{3\sqrt{x}}{x^2}$$

8. Simplify.

   (a) $\dfrac{\sqrt{5}}{\sqrt{24}}$

   (b) $\dfrac{5x}{\sqrt{x^9}}$

Example	Student Practice

9. Simplify. $\dfrac{\sqrt{2}}{\sqrt{27x}}$

Since it is not apparent what we should multiply 27 by to obtain a perfect square, we will begin by simplifying the denominator.

$\dfrac{\sqrt{2}}{\sqrt{27x}} = \dfrac{\sqrt{2}}{3\sqrt{3x}}$

$= \dfrac{\sqrt{2}}{3\sqrt{3x}} \times \dfrac{\sqrt{3x}}{\sqrt{3x}}$

$= \dfrac{\sqrt{6x}}{3\sqrt{9x^2}}$

$= \dfrac{\sqrt{6x}}{9x}$

10. Simplify. $\dfrac{\sqrt{3x}}{\sqrt{50x}}$

---

11. Simplify. $\dfrac{2}{\sqrt{3}-4}$

The conjugate of $\sqrt{3}-4$ is $\sqrt{3}+4$.

$\dfrac{2}{\sqrt{3}-4} \cdot \dfrac{\sqrt{3}+4}{\sqrt{3}+4}$

$= \dfrac{2\sqrt{3}+8}{\left(\sqrt{3}\right)^2 + 4\sqrt{3} - 4\sqrt{3} - 4^2}$

$= \dfrac{2\sqrt{3}+8}{3-16}$

$= \dfrac{2\sqrt{3}+8}{-13}$

$= -\dfrac{2\sqrt{3}+8}{13}$

12. Simplify.

(a) $\dfrac{6}{\sqrt{2}-\sqrt{7}}$

(b) $\dfrac{\sqrt{y}}{\sqrt{17}-4}$

Example	Student Practice
13. Rationalize the denominator. $\dfrac{\sqrt{3}+\sqrt{2}}{\sqrt{3}-\sqrt{2}}$	**14.** Rationalize the denominator. $\dfrac{\sqrt{6}+\sqrt{5}}{\sqrt{6}-\sqrt{5}}$

The conjugate of $\sqrt{3}-\sqrt{2}$ is $\sqrt{3}+\sqrt{2}$.

$$\frac{\sqrt{3}+\sqrt{2}}{\sqrt{3}-\sqrt{2}} \cdot \frac{\sqrt{3}+\sqrt{2}}{\sqrt{3}+\sqrt{2}}$$

$$= \frac{\sqrt{9}+\sqrt{6}+\sqrt{6}+\sqrt{4}}{\left(\sqrt{3}\right)^2 - \left(\sqrt{2}\right)^2}$$

$$= \frac{3+2\sqrt{6}+2}{3-2}$$

$$= \frac{5+2\sqrt{6}}{1}$$

$$= 5+2\sqrt{6}$$

Extra Practice

1. Simplify. Be sure to rationalize all denominators. Do not use a calculator or a table of square roots. $\dfrac{\sqrt{48}}{\sqrt{12}}$

2. Simplify. Be sure to rationalize all denominators. Do not use a calculator or a table of square roots. $\sqrt{\dfrac{3}{15}}$

3. Rationalize the denominator. Simplify your answer. Do not use a calculator or a table of square roots. $\dfrac{x+4}{\sqrt{x}-2}$

4. Rationalize the denominator. Simplify your answer. Do not use a calculator or a table of square roots. $\dfrac{3\sqrt{5}+2}{\sqrt{8}-\sqrt{6}}$

Concept Check

Explain how you would simplify $\dfrac{x-4}{\sqrt{x}+2}$.

Chapter 9 Radicals
9.6 The Pythagorean Theorem and Radical Equations

Vocabulary
hypotenuse • legs • Pythagorean Theorem • radical equation • extraneous root

1. A(n) _____ is an equation with a variable in one or more of the radicands.

2. An apparent solution that does not satisfy the original equation is called a(n)
    _____.

3. The shorter sides of a right triangle are referred to as the _____.

4. The longest side of a right triangle is called the _____.

Example	Student Practice
1. The ramp to Tony Pitkin's barn rises 5 feet over a horizontal distance of 12 feet. How long is the ramp? That is, find the length of the hypotenuse of a right triangle whose legs are 5 feet and 12 feet.	**2.** Find the length of the hypotenuse of a right triangle with legs of 24 yards and 32 yards.

Write the Pythagorean Theorem. Substitute the known values into the equation.

$$c^2 = a^2 + b^2$$
$$c^2 = 5^2 + 12^2$$
$$c^2 = 25 + 144$$
$$c^2 = 169$$
$$c = \pm\sqrt{169}$$
$$= \pm 13$$

The hypotenuse is 13 feet. We do not use -13 because length is not negative.

The check is left to the student.

Vocabulary Answers: 1. radical equation 2. extraneous root 3. legs 4. hypotenuse

Example	Student Practice
3. A 25-foot ladder is placed against a building. The foot of the ladder is 8 feet from the wall. At approximately what height does the top of the ladder touch the building? Round to the nearest tenth.	**4.** A kite is on a string that is 50 feet long and is fastened to the ground at the other end. Assuming the string is a straight line, how high is the kite if it is flying 8 feet away from where it is fastened to the ground? Round to the nearest tenth.

$$c^2 = a^2 + b^2$$
$$25^2 = 8^2 + b^2$$
$$625 = 64 + b^2$$
$$561 = b^2$$
$$\pm\sqrt{561} = b$$

We want only the positive value for the distance, so $b = \sqrt{561}$. Using a calculator, we have $561\ \boxed{\sqrt{\ }}\ 23.685438$. Rounding, we obtain $b \approx 23.7$. The ladder touches the building at a height of approximately 23.7 feet.

---

5. Solve and check. $1 + \sqrt{5x - 4} = 5$

$$1 + \sqrt{5x - 4} = 5$$
$$\sqrt{5x - 4} = 4$$
$$\left(\sqrt{5x - 4}\right)^2 = \left(4\right)^2$$
$$5x - 4 = 16$$
$$5x = 20$$
$$x = 4$$

Check.

$$1 + \sqrt{5(4) - 4} \overset{?}{=} 5$$
$$1 + \sqrt{20 - 4} \overset{?}{=} 5$$
$$1 + \sqrt{16} \overset{?}{=} 5$$
$$1 + 4 = 5$$

6. Solve and check. $-3 + \sqrt{2x - 5} = 6$

Example	Student Practice

7. Solve and check. $\sqrt{x+3} = -7$

$$\sqrt{x+3} = -7$$
$$x+3 = 49$$
$$x = 46$$

Check.

$$\sqrt{46+3} \overset{?}{=} -7$$
$$\sqrt{49} \overset{?}{=} -7$$
$$7 \neq -7$$

There is no solution.

8. $\sqrt{6x+8} = -3$

---

9. Solve and check. $2 + \sqrt{2x-1} = x$

$$2 + \sqrt{2x-1} = x$$
$$\left(\sqrt{2x-1}\right)^2 = (x-2)^2$$
$$2x-1 = x^2 - 4x + 4$$
$$0 = x^2 - 6x + 5$$
$$0 = (x-5)(x-1)$$
$$x = 5 \text{ or } x = 1$$

Check.
If $x = 5$:

$$2 + \sqrt{2(5)-1} \overset{?}{=} 5$$
$$\sqrt{9} \overset{?}{=} 3$$
$$3 = 3$$

If $x = 1$:

$$2 + \sqrt{2(1)-1} \overset{?}{=} 1$$
$$\sqrt{1} \overset{?}{=} -1$$
$$1 \neq -1$$

Thus only 5 is a solution to this equation.

10. Solve and check. $\sqrt{4x+1} + 3 = x - 2$

Example	Student Practice
11. Solve and check. $\sqrt{3x+3} = \sqrt{5x-1}$	**12.** Solve and check. $\sqrt{3x+4} = \sqrt{2x+8}$

Here there are two radicals. Each radical is already isolated.

$$\sqrt{3x+3} = \sqrt{5x-1}$$
$$\left(\sqrt{3x+3}\right)^2 = \left(\sqrt{5x-1}\right)^2$$
$$3x+3 = 5x-1$$
$$3 = 2x-1$$
$$4 = 2x$$
$$2 = x$$

Check.

$$\sqrt{3(2)+3} \overset{?}{=} \sqrt{5(2)-1}$$
$$\sqrt{6+3} \overset{?}{=} \sqrt{10-1}$$
$$\sqrt{9} = \sqrt{9}$$

Extra Practice

1. A right triangle has legs a and b and hypotenuse c. Find the exact length of the missing side c if $a = 9$ and $b = 9$.

2. A right triangle has legs a and b and hypotenuse c. Find the exact length of the missing side b if $a = \sqrt{72}$ and $c = 10$.

3. Solve for the variable. Check your solutions. $\sqrt{2x+5} = 7$

4. Solve for the variable. Check your solutions. $\sqrt{4x-13} = 2x-6$

Concept Check

A student attempted to solve the equation $2 + \sqrt{1-8x} = x$ and found values of $x = -3$ and $x = -1$. Explain how you would check these values to see if they are solutions to the original equation.

Chapter 9 Radicals
9.7 Word Problems Involving Radials: Direct and Inverse Variation

Vocabulary
varies directly • constant of variation • vary inversely

1. The constant is often called the _____.

2. If y _____ as x, then $y = kx$, where k is a constant.

3. If one variable is a constant multiple of the reciprocal of another, the two variables are said to _____.

Example	Student Practice
1. Cliff's salary varies directly as the number of hours worked. Last week he earned $33.60 for working 7 hours. This week he earned $52.80. How many hours did he work?	**2.** Mary's salary varies directly as the number of hours worked. Last week she earned $41.84 for working 8 hours. This week she earned $78.45. How many hours did she work?

Let S = salary, h = number of hours worked, and k = constant of variation. Since his salary varies directly as the number of hours he worked, $S = k \cdot h$. Find the constant k by substituting the known values of $S = 33.60$ and $h = 7$.

$$33.60 = 7k$$
$$\frac{33.60}{7} = \frac{7k}{7}$$
$$4.80 = k$$

How many hours did he work to earn $52.80?

$$S = 4.80h$$
$$52.80 = 4.80h$$
$$11 = h$$

Cliff worked 11 hours this week.

Vocabulary Answers: 1. constant of variation 2. varies directly 3. vary indirectly

Example	Student Practice

3. In a certain class of racing cars, the maximum speed varies directly as the square root of the horsepower of the engine. If a car with 225 horsepower can achieve a maximum speed of 120 mph, what speed could it achieve with 256 horsepower?

Let $V =$ the maximum speed, $h =$ the horsepower of the engine, and $k =$ the constant of variation.

Since the maximum speed (V) varies directly as the square root of the horsepower of the engine, we can write the following.

$$V = k\sqrt{h}$$
$$120 = k\sqrt{225}$$
$$120 = 15k$$
$$8 = k$$

Now we can write the direct variation equation with the known value for k.

$$V = 8\sqrt{h}$$
$$= 8\sqrt{256}$$
$$= (8)(16)$$
$$= 128$$

Thus, a car with 256 horsepower could achieve a maximum speed of 128 mph.

4. The volume of a certain type of balloon varies directly as the cube of the amount of air put in the balloon. If the volume was 32 centimeters cubed when 2 milliliters of air was in the balloon, what is the volume when 7 milliliters of air is in the balloon?

Example	Student Practice
5. A car manufacturer is thinking of reducing the size of the wheel used in a subcompact car. The number of times a car wheel must turn to cover a given distance varies inversely as the radius of the wheel. (Notice that this says that the smaller the wheel, the more times it must turn to cover a given distance.) A wheel with a radius of 0.35 meter must turn 400 times to cover a specified distance on a test track. How many times would it have to turn if the radius were reduced to 0.30 meter?	**6.** A car manufacturer is thinking of reducing the size of the wheel used in a subcompact car. The number of times a car wheel must turn to cover a given distance varies inversely as the radius of the wheel. (Notice that this says that the smaller the wheel, the more times it must turn to cover a given distance.) A wheel with a radius of 0.40 meter must turn 600 times to cover a specified distance on a test track. How many times would it have to turn if the radius were reduced to 0.32 meter?

Let n = the number of times the car wheel turns, r = the radius of the wheel, and k = the constant of variation. Since the number of turns varies inversely as the radius, we can write the following.

$$n = \frac{k}{r}$$

$$400 = \frac{k}{0.35}$$

$$140 = k$$

How many times must the wheel turn if the radius is 0.30 meter?

$$n = \frac{140}{r}$$

$$n = \frac{140}{0.30}$$

$$n = 466\frac{2}{3}$$

The wheel would have to turn $466\frac{2}{3}$ times to cover the same distance if the radius were only 0.30 meter.

Example	Student Practice
7. The illumination of a light source varies inversely as the square of the distance from the source. The illumination measures 25 candlepower when a certain light is 4 meters away. Find the illumination when the light is 8 meters away. Since the illumination varies inversely as the square of the distance, $I = \dfrac{k}{d^2}$. $$I = \frac{k}{d^2}$$ $$25 = \frac{k}{4^2}$$ $$400 = k$$ $$I = \frac{400}{d^2} = \frac{400}{8^2} = \frac{25}{4} = 6.25$$ The illumination is 6.25 candlepower when the light source is 8 meters away.	**8.** The resistance to the flow of electric current in a wire of fixed length varies inversely as the square of the diameter of the wire. When the resistance measured is 0.45 ohm, the diameter of the wire is 0.02 centimeter. If we use wire that is 0.04 centimeter in diameter, what will the resistance be?

Extra Practice

1. If y varies directly as the cube of x, and $y = 3$ when $x = \dfrac{1}{2}$, find y when $x = 4$.

2. If y varies inversely as the square of x, and $y = 3$ when $x = \dfrac{1}{3}$, find y when $x = 3$.

3. The cost of shipping a package varies directly as the weight of the package. If the cost of shipping a 15 pound package is $11.25, what is the cost of shipping a package that weighs 180 pounds?

4. The amount of time in minutes it takes for an ice cube to melt varies inversely with the temperature of the water that the ice cube is placed in. When an ice cube is placed in 45°F water, it takes 4.3 minutes to melt. How long would it take an ice cube of this size to melt if it were placed in 75°F water?

Concept Check

The distance a body falls from rest varies directly as the square of the time it falls. If an object falls 64 feet in two seconds, explain how you would write an equation to describe this relationship. How would you find the constant k in this equation?

274

MATH COACH

Mastering the skills you need to do well on the test.

Watch the **MATH COACH** videos in MyMathLab®or on YouTube™ while you work the problems below. These helpful hints will help you avoid making common errors on test problems.

Adding Radical Expressions That Require Simplifying—

Problem 6   Combine and simplify. $\sqrt{4a} + \sqrt{8a} + \sqrt{36a} + \sqrt{18a}$

> **Helpful Hint:** You must simplify each radical first. Terms can only be combined when the expression inside the radical is exactly the same.

Did you identify that 4 and 36 are perfect squares and then simplify the first radical to $2\sqrt{a}$ and the third radical to $6\sqrt{a}$? Yes _____   No _____

If you answered No, notice that the two radicals can be rewritten as $\sqrt{4} \cdot \sqrt{a}$ and $\sqrt{36} \cdot \sqrt{a}$. You can then take the square root of 4 and the square root of 36. Go back and complete this step again.

Did you look for perfect squares and then rewrite the second radical as $\sqrt{4 \cdot 2 \cdot a}$ and the fourth radical as $\sqrt{9 \cdot 2 \cdot a}$?

Yes _____   No _____

If you answered No, remember that we want to factor the radicand into products of perfect squares whenever possible. Now the square roots of 4 and 9 can be taken to simplify these radicals further.

The last step is to combine any radicals where the expression inside the radical is exactly the same.

If you answered Problem 6 incorrectly, go back and rework the problem using these suggestions.

Multiplying Two Binomial Radical Expressions—Problem 10   Multiply $\left(4\sqrt{2} - \sqrt{5}\right)\left(3\sqrt{2} + \sqrt{5}\right)$.

> **Helpful Hint:** Remember that you can use the FOIL method to multiply any two binomials. This also applies to binomial radical expressions. Be careful to separate numbers outside the radical and numbers inside the radical when completing the multiplication.

When you multiplied the first two terms and simplified that product, did you get 24?
Yes _____   No _____

If you answered No, remember we must separately multiply the numbers outside the radicals and then multiply the numbers inside the radical signs. Be sure to simplify radical expressions such as $\sqrt{4}$.

When you multiplied the outer terms, did you get $4\sqrt{10}$?
Yes _____   No _____

If you answered No, notice that $4\sqrt{2} \cdot \sqrt{5} = 4 \cdot 1\sqrt{2 \cdot 5}$
$= 4\sqrt{10}$.

The next steps are to multiply the inner terms, multiply the last two terms, and then combine any like terms.

Now go back and rework the problem using these suggestions.

Simplifying the Quotient of Two Binomials That Contain Radicals—Problem 13

Simplify. $\dfrac{\sqrt{3}+4}{5+\sqrt{3}}$

> **Helpful Hint:** When the denominator of a fraction contains a radical expression, we multiply both the numerator and the denominator of the fraction by the conjugate of the denominator.

Did you identify the conjugate of the denominator as $5-\sqrt{3}$? Yes ____ No ____

Did you multiply the numerator and denominator by $5-\sqrt{3}$ to obtain the expression $\dfrac{\left(\sqrt{3}+4\right)\left(5-\sqrt{3}\right)}{\left(5+\sqrt{3}\right)\left(5-\sqrt{3}\right)}$?

Yes ____ No ____

If you answered No to these questions, go back and review the definition of conjugate and try these two steps again.

After completing the multiplication, did you get the unsimplified expression $\dfrac{5\sqrt{3}-\sqrt{9}+20-4\sqrt{3}}{25-5\sqrt{3}+5\sqrt{3}-\sqrt{9}}$?

Yes ____ No ____

If you answered No, slowly go through the step of multiplying the two binomials in the numerator.

Then carefully multiply the two binomials in the denominator.

After completing these steps, you can look for the square roots to evaluate. Then combine any like terms separately in both the numerator and denominator. Your final result will still be a fraction.

If you answered Problem 13 incorrectly, go back and rework the problem using these suggestions.

Solving and Verifying the Solution of a Radical Equation—Problem 19

Solve. Verify your solutions. $x = 5 + \sqrt{x+7}$

> **Helpful Hint:** Always isolate the radical on one side of the equation before squaring each side of the equation. Be sure to check your results for extraneous roots.

Did you first isolate the radical to get $x-5 = \sqrt{x+7}$?
Yes ____ No ____

When you squared each side of this equation, did you multiply and then simplify to obtain the equation $x^2 - 10x + 25 = x + 7$? Yes ____ No ____

If you answered No to these questions, notice that $(x-5)^2 = (x-5)(x+5)$ on the left side. Also remember that $\left(\sqrt{x+7}\right)^2 = \sqrt{x+7} \cdot \sqrt{x+7} = x+7$.
Stop now and complete these calculations.

Did you transform the equation to $x^2 - 11x + 18 = 0$?
Yes ____ No ____

If you answered No, try to get all the terms on the left and only 0 on the right. Then factor the resulting quadratic equation.

When you check both possible answers in the original equation, did both answers work?
Yes ____ No ____

If you answered Yes, carefully replace x by 2 in the equation. Remember that when you obtain $2 = 5 + 3$ you can see immediately that this is an invalid equation. The value $x = 2$ does not check. It is an extraneous root.

Now go back and rework the problem using these suggestions.

Chapter 10 Quadratic Equations
10.1 Introduction to Quadratic Equations

Vocabulary
quadratic equation • standard form • greatest common factor • zero factor property

1. The _____ of a quadratic equation is $ax^2 + bx + c = 0$, where a, b, and c are real numbers and $a \neq 0$.

2. If it is possible to factor the quadratic equation, then we use the _____ to find the solution.

3. To solve equations of the form $ax^2 + bx = 0$, begin by factoring out the _____ .

4. A _____ is a polynomial of degree two.

Example	Student Practice
1. Place each quadratic equation in standard form and identify the real numbers a, b, and c.	**2.** Place each quadratic equation in standard form and identify the real numbers a, b, and c.
(a) $5x^2 - 6x + 3 = 0$	**(a)** $3x^2 + 4x - 6 = 0$
This equation is in standard form, $ax^2 + bx + c = 0$. Match each term to the standard form. $5x^2 - 6x + 3 = 0$ $a = 5$, $b = -6$, $c = 3$	**(b)** $5x + 7x^2 - 6 = 0$
(b) $-2x^2 + 15x + 4 = 0$	**(c)** $4x^2 - 7x = -3$
It is easier to work with quadratic equations if the first term is not negative. Multiply each term on both sides of the equation by -1.	**(d)** $6x^2 + 3 = 0$
$2x^2 - 15x - 4 = 0$ $a = 2$, $b = -15$, $c = -4$	

Vocabulary Answers: 1. standard form 2. zero factor property 3. greatest common factor 4. quadratic equation

Example	Student Practice
3. Solve. $7x^2 + 9x - 2 = -8x - 2$	**4.** Solve. $3x^2 - 9x + 7 = 7 - 6x$

3. Solve. $7x^2 + 9x - 2 = -8x - 2$

The equation is not in standard form.
Add $8x + 2$ to each side and simplify.
$$7x^2 + 9x - 2 = -8x - 2$$
$$7x^2 + 9x - 2 + 8x + 2 = 0$$
$$7x^2 + 17x = 0$$

Factor. Then set each factor equal to zero and solve for x.

$$7x^2 + 17x = 0$$
$$x(7x + 17) = 0$$
$$x = 0 \qquad 7x + 17 = 0$$
$$\qquad\qquad 7x = -17$$
$$\qquad\qquad x = -\frac{17}{7}$$

The solutions are 0 and $-\dfrac{17}{7}$.

4. Solve. $3x^2 - 9x + 7 = 7 - 6x$

---

5. Solve and check. $8x - 6 + \dfrac{1}{x} = 0$

Multiply each term by the LCD, x.
$$x(8x) - x(6) + x\left(\frac{1}{x}\right) = x(0)$$
$$8x^2 - 6x + 1 = 0$$

Factor the equation and solve for x.
$$8x^2 - 6x + 1 = 0$$
$$(4x - 1)(2x - 1) = 0$$
$$4x - 1 = 0 \qquad 2x - 1 = 0$$
$$x = \frac{1}{4} \qquad\qquad x = \frac{1}{2}$$
The check is left to the student.

6. Solve and check. $12x + 7 + \dfrac{1}{2x} = 0$

Example	Student Practice
7. A truck delivery company can handle a maximum of 36 truck routes in one day, and every two cities have a distinct truck route between them. How many separate cities can the truck company service? Use the equation $t = \dfrac{n^2 - n}{2}$.	**8.** A truck delivery company can handle a maximum of 21 truck routes in one day, and every two cities have a distinct truck route between them. How many separate cities can the truck company service? Use the equation $t = \dfrac{n^2 - n}{2}$.

Substitute 36 for t, the number of truck routes.

$$t = \frac{n^2 - n}{2}$$

$$36 = \frac{n^2 - n}{2}$$

Multiply both sides by 2. Then write the equation in standard form.

$$36 = \frac{n^2 - n}{2}$$

$$2(36) = 2\left(\frac{n^2 - n}{2}\right)$$

$$72 = n^2 - n$$

$$0 = n^2 - n - 72$$

Factor.

$$0 = n^2 - n - 72$$

$$0 = (n - 9)(n + 8)$$

Set each factor equal to zero and solve.

$$n - 9 = 0 \qquad n + 8 = 0$$

$$n = 9 \qquad\quad n = -8$$

We reject -8 as a meaningless solution, because we cannot have a negative number of cities. Thus, the trucking company can service 9 cities.

Extra Practice

1. Place the quadratic equation in standard form and identify the real numbers a, b, and c.

   $2x^2 - 27 = -3x^2 + 12$

2. Solve. $6x^2 + 3x = 0$

3. Solve and check. $x^2 = 10x - 21$

4. Solve and check. $\dfrac{8}{x} = \dfrac{x}{3} + \dfrac{5}{3}$

Concept Check

In the following problem, explain how you would place the quadratic equation in standard form. $2 = \dfrac{5}{x+1} + \dfrac{3}{x-1}$

Name: _____   Date: _____
Instructor: _____   Section: _____

Chapter 10 Quadratic Equations
10.2 Using the Square Root Property and Completing the Square to Find Solutions

Vocabulary
square root property   •   completing the square

1. Rewriting an equation so that it has the form $(ax+b)^2 = c$ is called _____.

2. The _____ states that if $x^2 = a$, then $x = \sqrt{a}$ or $x = -\sqrt{a}$, for all nonnegative real numbers a.

Example	**Student Practice**
1. Solve.	**2.** Solve.
(a) $x^2 = 49$	**(a)** $x^2 = 100$
$x^2 = 49$ $x = \pm\sqrt{49}$ $x = \pm 7$	
(b) $x^2 = 20$	**(b)** $x^2 = 72$
$x^2 = 20$ $x = \pm\sqrt{20}$ $x = \pm 2\sqrt{5}$	
(c) $5x^2 = 125$	**(c)** $6x^2 = 120$
Divide both sides by 5 before taking the square root. $\dfrac{5x^2}{5} = \dfrac{125}{5}$ $x^2 = 25$ $x = \pm\sqrt{25}$ $x = \pm 5$	

Vocabulary Answers: 1. completing the square 2 square root property

Example	Student Practice
3. Solve. $3x^2 + 5x = 18 + 5x + x^2$	**4.** Solve. $5x^2 + 2x - 13 = 2x + 35 + 2x^2$

Simplify the equation by placing all the variable terms on the left and the constants on the right.

$$3x^2 + 5x = 18 + 5x + x^2$$
$$2x^2 = 18$$
$$x^2 = 9$$
$$x = \pm 3$$

---

5. Solve. $(3x+1)^2 = 8$	**6.** Solve. $(2x-5)^2 = 20$

Take the square root of both sides and simplify as much as possible.

$$(3x+1)^2 = 8$$
$$3x+1 = \pm\sqrt{8}$$
$$3x+1 = \pm 2\sqrt{2}$$

Now we must solve the two equations expressed by the plus or minus statement.

$$3x+1 = +2\sqrt{2} \qquad 3x+1 = -2\sqrt{2}$$
$$3x = -1 + 2\sqrt{2} \qquad 3x = -1 - 2\sqrt{2}$$
$$x = \frac{-1+2\sqrt{2}}{3} \qquad x = \frac{-1-2\sqrt{2}}{3}$$

The roots of this quadratic equation are irrational numbers. They are
$$\frac{-1+2\sqrt{2}}{3} \text{ and } \frac{-1-2\sqrt{2}}{3} \text{ or } \frac{-1\pm2\sqrt{2}}{3}.$$
We cannot simplify these roots further, so we leave them in this form.

Example	Student Practice
7. Solve. $4x^2 + 4x - 3 = 0$	**8.** Solve. $3x^2 - 6x - 24 = 0$

7. Solve. $4x^2 + 4x - 3 = 0$

Place the constant on the right. This puts the equation in the form $ax^2 + bx = c$. Then divide all terms by 4.

$$4x^2 + 4x - 3 = 0$$
$$4x^2 + 4x = 3$$
$$x^2 + x = \frac{3}{4}$$

Take one-half the coefficient of x and square it, $\left(\frac{1}{2}\right)^2 = \frac{1}{4}$. Add $\frac{1}{4}$ to each side.

$$x^2 + x + \frac{1}{4} = \frac{3}{4} + \frac{1}{4}$$

Factor the left side. Then use the square root property to solve for x.

$$x^2 + x + \frac{1}{4} = \frac{3}{4} + \frac{1}{4}$$

$$\left(x + \frac{1}{2}\right)^2 = 1$$

$$x + \frac{1}{2} = \pm\sqrt{1}$$

$$x + \frac{1}{2} = \pm 1$$

$$x + \frac{1}{2} = 1 \qquad x + \frac{1}{2} = -1$$

$$x = \frac{1}{2} \qquad x = -\frac{3}{2}$$

Thus the two roots are $\frac{1}{2}$ and $-\frac{3}{2}$.

The check is left to the student.

Extra Practice

1. Solve using the square root property.

   $3x^2 = 192$

2. Solve using the square root property.

   $(2x-4)^2 = 60$

3. Solve by completing the square.

   $x^2 - 8x = 11$

4. Solve by completing the square.

   $2x^2 - 5x - 4 = 0$

Concept Check

Explain the first three steps of how you would solve the following problem by completing the square. $5x^2 - 10x + 2 = 0$

Chapter 10 Quadratic Equations
10.3 Using the Quadratic Formula to Find Solutions

Vocabulary
quadratic formula   •   discriminant

1. If the _____ is a negative number, the roots are not real numbers.

2. Use the _____ to find the roots of any quadratic equation of the form $ax^2 + bx + c = 0$, where a, b, and c are real numbers and $a \neq 0$.

Example	Student Practice
1. Solve using the quadratic formula. $3x^2 + 10x + 7 = 0$ In our given equation, we have $a = 3$, $b = 10$, and $c = 7$. Write the quadratic formula and substitute the values for a, b, and c. Then, simplify.	**2.** Solve using the quadratic formula. $2x^2 + 3x - 5 = 0$

$$x = \frac{-b \pm \sqrt{b^2 - 4ac}}{2a}$$

$$= \frac{-10 \pm \sqrt{(10)^2 - 4(3)(7)}}{2(3)}$$

$$= \frac{-10 \pm \sqrt{100 - 84}}{6}$$

$$= \frac{-10 \pm \sqrt{16}}{6} = \frac{-10 \pm 4}{6}$$

Solve using the positive sign.
$$x = \frac{-10 + 4}{6} = \frac{-6}{6} = -1$$

Solve using the negative sign.
$$x = \frac{-10 - 4}{6} = \frac{-14}{6} = -\frac{7}{3}$$

Vocabulary Answers: 1. discriminant 2 quadratic formula

Example	Student Practice

3. Solve. $x^2 = 5 - \dfrac{3}{4}x$

First, we obtain an equivalent equation that does not have fractions. Multiply each term by the LCD, 4, and simplify.

$$x^2 = 5 - \frac{3}{4}x$$

$$4(x^2) = 4(5) - 4\left(\frac{3}{4}x\right)$$

$$4x^2 = 20 - 3x$$

Add $3x - 20$ to each side to write the equation in standard form.

$$4x^2 = 20 - 3x$$

$$4x^2 + 3x - 20 = 0$$

$a = 4, \ b = 3, \ \text{and} \ c = -20$

Substitute the values for a, b, and c, into the quadratic formula.

$$x = \frac{-b \pm \sqrt{b^2 - 4ac}}{2a}$$

$$x = \frac{-3 \pm \sqrt{(3)^2 - 4(4)(-20)}}{2(4)}$$

$$x = \frac{-3 \pm \sqrt{9 + 320}}{8}$$

$$x = \frac{-3 \pm \sqrt{329}}{8}$$

4. Solve. $x^2 = \dfrac{2}{3}x + 4$

Example	Student Practice

5. Find the roots of $3x^2 - 5x = 7$. Approximate to the nearest thousandth.

Place the equation in standard form.

$$3x^2 - 5x = 7$$
$$3x^2 - 5x - 7 = 0$$

$a = 3$, $b = -5$, and $c = -7$

Substitute the values for a, b, and c into the quadratic formula, then simplify.

$$x = \frac{-b \pm \sqrt{b^2 - 4ac}}{2a}$$

$$x = \frac{-(-5) \pm \sqrt{(-5)^2 - 4(3)(-7)}}{2(3)}$$

$$x = \frac{5 \pm \sqrt{25 + 84}}{6}$$

$$x = \frac{5 \pm \sqrt{109}}{6}$$

Look up $\sqrt{109}$ in the square root table. $\sqrt{109} \approx 10.440$. Use this result to solve for x.

$$x \approx \frac{5 + 10.440}{6} = \frac{15.440}{6} \approx 2.573$$

$$x \approx \frac{5 - 10.440}{6} = \frac{-5.440}{6} \approx -0.907$$

The two roots are 2.573 and -0.907.

6. Find the roots of $4x^2 - 7x = 12$. Approximate to the nearest thousandth.

Example	Student Practice

7. Determine whether $3x^2 = 5x - 4$ has real number solution(s).

First, we place the equation in standard form.
$$3x^2 = 5x - 4$$
$3x^2 - 5x + 4 = 0$
$a = 3,\ b = -5,\ c = 4$

Substitute the values for a, b, and c into the discriminant, $b^2 - 4ac$.

$$b^2 - 4ac = (-5)^2 - 4(3)(4)$$
$$= 25 - 48$$
$$= -23$$

The discriminant is negative. Thus $3x^2 = 5x - 4$ has no real number solution(s).

8. Determine whether $4x^2 = 6x - 5$ has real number solution(s).

Extra Practice

1. Solve. $x^2 - 2x = 4$

2. Solve. $3x^2 - 2x + 4 = 0$

3. Solve. $\dfrac{1}{5} + \dfrac{4}{x} = \dfrac{x}{5}$

4. Solve. $3x^2 + 8x + 12 = 0$

Concept Check

Explain how you would determine if the quadratic equation $5x^2 - 8x + 9 = 0$ has real solutions or has no real solutions?

Name: _____ Date: _____

Instructor: _____ Section: _____

Chapter 10 Quadratic Equations
10.4 Graphing Quadratic Equations

Vocabulary

parabola   •   vertex

1. The _____ is the highest point of a parabola that opens downward or the lowest point of a parabola that opens upward.

2. The graph of a quadratic equation is shaped like a _____.

Example	**Student Practice**
1. Graph.	**2.** Graph.

1. Graph.

(a) $y = x^2$

Make a table of values.

x	y
−3	9
−2	4
−1	1
0	0
1	1
2	4
3	9

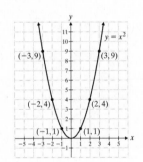

(b) $y = -x^2$

x	y
−3	−9
−2	−4
−1	−1
0	0
1	−1
2	−4
3	−9

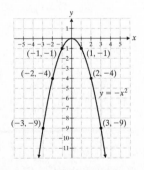

Notice that when the coefficient of x^2 is negative, the graph flips upside down.

2. Graph.

(a) $y = \frac{1}{4}x^2$

x	y
4	
2	
0	
−2	
−4	

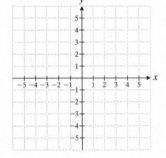

(b) $y = -\frac{1}{4}x^2$

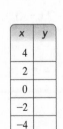

x	y
4	
2	
0	
−2	
−4	

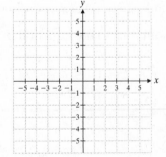

Vocabulary Answers: 1. vertex 2. parabola

Example	Student Practice

Example

3. Graph $y = x^2 - 2x$. Identify the coordinates of the vertex.

First, make a table of values.

x	y
−2	8
−1	3
0	0
1	−1
2	0
3	3
4	8

Next, plot the points.

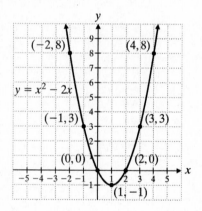

Notice that the x-intercepts are at $(0,0)$ and $(2,0)$. The x-coordinates are the solutions to the equation $x^2 - 2x = 0$.

$$x^2 - 2x = 0$$
$$x(x-2) = 0$$
$$x = 0, \quad x = 2$$

Student Practice

4. Graph $y = x^2 + 4x$. Identify the coordinates of the vertex.

x	y
1	
0	
−1	
−2	
−3	
−4	
−5	

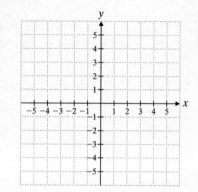

Example	Student Practice
5. $y = -x^2 - 2x + 3$. Determine the vertex and the x-intercepts. Then sketch the graph.	**6.** $y = x^2 - 6x + 5$. Determine the vertex and the x-intercepts, then sketch the graph.

We will first determine the coordinates of the vertex. We begin by finding the x-coordinate.

$$x = \frac{-b}{2a} = \frac{-(-2)}{2(-1)} = \frac{2}{-2} = -1$$

$$y = -x^2 - 2x + 3$$
$$y = -(-1)^2 - 2(-1) + 3 = 4$$

The point $(-1, 4)$ is the vertex.

Find the x-intercepts, the solutions to the equation $-x^2 - 2x + 3 = 0$. Since it is easier to factor if the first term is positive, add $x^2 + 2x - 3$ to each side. Then factor and solve for x.

$$-x^2 - 2x + 3 = 0$$
$$0 = x^2 + 2x - 3$$
$$0 = (x - 1)(x + 3)$$
$$x = 1 \qquad x = -3$$

The x-intercepts are $(1, 0)$ and $(-3, 0)$.

Graph. Since $a = -1$ the parabola opens down and the vertex is the highest point.

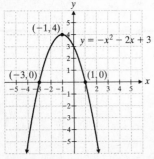

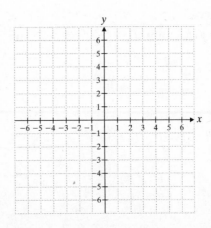

Extra Practice

1. Graph $y = x^2 - 3$.

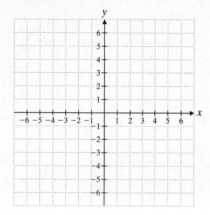

2. Graph $y = 3(x+1)^2$.

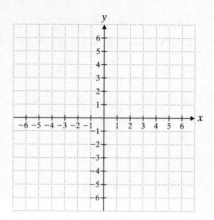

3. Graph $y = -x^2 + 2x + 4$.

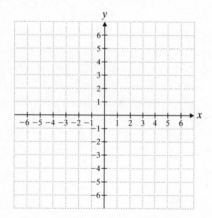

4. Graph $y = x^2 + 3x + 4$. Identify the coordinates of the vertex.

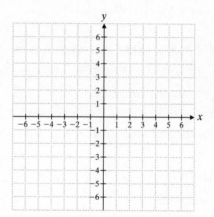

Concept Check

Explain how you would find the vertex of the equation $y = 4x^2 + 16x - 2$.

Chapter 10 Quadratic Equations
10.5 Solving Applied Problems Using Quadratic Equations

Vocabulary
quadratic equation   •   substitution

1. If there are two equations in two variables use _____ to solve for one variable, then use that result in the other equation to solve for the other variable.

2. The Pythagorean Theorem is a _____

Example	Student Practice
1. The hypotenuse of a right triangle is 25 meters in length. One leg is 17 meters longer than the other. Find the length of each leg.	**2.** The hypotenuse of a right triangle is 30 meters in length. One leg is 6 meters shorter than the other. Find the length of each leg.

Example

1. The hypotenuse of a right triangle is 25 meters in length. One leg is 17 meters longer than the other. Find the length of each leg.

Understand the problem. Draw a picture.

Since the problem involves a right triangle, use the Pythagorean Theorem.

$$a^2 + b^2 = c^2$$

$$x^2 + (x+17)^2 = 25^2$$

Solve and state the answer.

$$x^2 + (x+17)^2 = 25^2$$

$$x^2 + x^2 + 34x + 289 = 625$$

$$x^2 + 17x - 168 = 0$$

$$(x+24)(x-7) = 0$$

$$x + 24 = 0 \qquad x - 7 = 0$$

$$x = -24 \qquad x = 7$$

Note $x = -24$ is not a valid solution. Thus, one leg is 7 meters and the other leg is $7 + 17 = 24$ meters in length. The check is left to the student.

Vocabulary Answers: 1. substitution 2 quadratic equation

Example	Student Practice

3. The ski club rented a bus to travel to Mount Snow. The members agreed to share the cost of $180 equally. On the day of the ski trip, three members were sick with the flu and could not go. This increased the share of each person going on the trip by $10. How many people originally planned to attend?

Let s = the number of students and c = the cost for each student in the original group. Write an equation.

$s \cdot c = 180$ (1)

If three people were sick, the number of students dropped by three, but the cost for each increased by $10. The total is still $180

$(s-3)(c+10) = 180$ (2)

Solve equation (1) for c to get $c = \dfrac{180}{s}$.

Substitute the result into equation (2).

$$(s-3)(c+10) = 180$$

$$(s-3)\left(\frac{180}{s}+10\right) = 180$$

$$(s-3)\left(\frac{180+10s}{s}\right) = 180$$

$$(s-3)(180+10s) = 180s$$

$$10s^2 + 150s - 540 = 180s$$

$$10s^2 - 30s - 540 = 0$$

$$s^2 - 3s - 54 = 0$$

$$(s-9)(s+6) = 0$$

$$s = 9, \qquad s = -6$$

The number of students originally going on the ski trip was 9.

4. A group of friends wants to charter a fishing boat for a day. The fee for the boat is $750 per day. On the day of the charter, 5 of the friends cancelled, and the cost for each person going on the boat went up $25. How many friends were originally supposed to go?